Rethinking Technical Cooperation

*Reforms for
Capacity Building in Africa*

Regional Bureau for Africa,
United Nations Development Programme and
Development Alternatives, Inc.

Elliot J. Berg, Coordinator

Elliot Berg is Vice-President for Policy and Research at Development Alternatives, Inc. and an Adjunct Professor at the University of the Auvergne in France. He was for many years Professor of Economics at the University of Michigan and has been a consultant to governments and such agencies as the United Nations, World Bank, and International Monetary Fund.

The opinions in this publication are those of the contributors and do not necessarily reflect the views or policies of the United Nations Development Programme.

All rights reserved. No part of this publication may be reproduced or transmitted in any form or by any means, electronic or mechanical, including photocopy, recording, or any information storage and retrieval system, without permission in writing from the publisher.

Copyright © 1993 by United Nations Development Programme

Published in 1993 by United Nations Development Programme, 1 United Nations Plaza, New York, New York 10017

A CIP catalog record for this book is available from the Library of Congress
United Nations Sales # E.91.III.B.4
ISBN 92-1-126022-1

CONTENTS

Foreword   v
Acknowledgments   ix

1   TECHNICAL COOPERATION UNDER FIRE   1
    Recipient Country Viewpoint   4
    Donor Criticisms   10
    Formal Evaluations   15

2   DEFINITIONS, HISTORY, STATISTICS   41
    Concepts and Definitions   42
    Some Historical Background   64
    Quantitative Trends   70

3   IMPROVING THE DELIVERY OF TECHNICAL COOPERATION   93
    Failures in Technical Cooperation Delivery Systems   93
    Proposals for Reform   106
    Concluding Remarks   121

4   STRENGTHENING RECIPIENT MANAGEMENT   127
    The Context: Diffuse Public Sector Decision Making   129
    Donor Dominance of Technical Cooperation Management   133
    Costs of the Present Arrangements and Why They Persist   135
    The Need for Reform   139
    Getting From Here to There: Present and Proposed Approaches   147
    Toward a Gradualist Reform Strategy   156

5	**CREATING A MORE EFFECTIVE MARKET FOR TECHNICAL COOPERATION**	**165**
	Imperfections in the Market	167
	Introducing Market Elements	175
	Pitfalls to the Market Approach	188
6	**GETTING THE WORK ENVIRONMENT RIGHT**	**195**
	Public Sector Organizational Disarray and Low Salaries	199
	Donor and Government Responses	209
	What Should be Done	220
7	**SUMMARY AND CONCLUSIONS**	**243**
	The Attack on Technical Cooperation	244
	Sources of Ineffectiveness	246
	Consensus Reform Proposals	247
	Problems with the Consensus Reform Proposals	254
	Additional (Nonconsensus) Recommendations	263

ANNEX: STATISTICS ON TECHNICAL COOPERATION	**273**
Bibliography	301
Index	321

TABLES

2.1	Expatriate Employment as a Percentage of Total Employment of Trained Manpower, c 1967	66
2.2	Higher-Level Students Studying Abroad	67
2.3	Geographic Distribution of Technical Assistance Personnel from OECD Countries Serving Overseas in 1963	69

2.4	Trends in Technical Cooperation Grants to Sub-Saharan Africa, Real and Nominal	73
2.5	Technical Cooperation Personnel by Category, Selected Countries, 1988 and 1989	75
2.6	Specialization of Technical Assistance Personnel, Selected Countries, 1989	78
2.7	Trends in Technical Cooperation Grant Disbursements from Major Donor Sources, 1970-1989	80
2.8	Major Technical Cooperation Recipients, 1989	83
2.9	Type of Technical Cooperation Expenditures, Selected Countries, c 1989	85
6-1	Wage and Employment Changes, 1957-1983, Selected Countries	205
6-2	1985 Base Civil Service Salary Rate as Percentage of 1975 Rate	206
6-3	Skill Differential Ratios for Selected Countries	208
6-4	Employment Reduction Mechanisms and Results for Selected Countries, 1981-1990	216

FIGURES

1	Trends in Technical Cooperation Grants	74
2	Technical Cooperation Grants as a Part of Official Development Assistance	74

FOREWORD

Nothing counts for more in the development of nations than sound economic policy and efficient economic management. How sound policies are defined varies among countries and over time. And efficiency in the management of national resources does not mean only the pursuit of increased output. But, in the end, poor countries and poor people need economic growth; without it, poverty endures and human potential continues to be wasted.

Improved policy making and better economic management—and self-reliance in these matters—are the central objectives of technical cooperation. For a long time, and even to a reduced extent today, this has meant educating people from developing countries in the universities of the industrialized world, helping in the building of local academic institutions, and sending expatriate technical assistance to train national staff on the job and to fill gaps in local skills.

In Sub-Saharan Africa, massive efforts have been made to expand education since independence some 30 years ago. The stock and flow of educated manpower have now reached substantial levels. Combined with other factors, this increase in educated manpower has led to a change in emphasis in technical cooperation. Much higher priority is now being given to capacity building—the enhancement of skills and the strengthening of institutions. But at the same time that capacity-building objectives have moved to center stage, practitioners and observers concerned with technical cooperation have become increasingly unsure about the ability of existing approaches to technical cooperation to meet these new priorities.

These uncertainties have profound implications for Sub-Saharan Africa. The region receives large amounts of assistance in the form of technical cooperation—$3.2 billion in grants in 1989, and significantly more when loans and uncounted transfers are included. This amounts to over a quarter of the total official development assistance that flows to Sub-Saharan Africa. So technical cooperation is a major economic variable, and the effectiveness with which it is used is a matter of deep importance in macroeconomic terms. The shrinking of overall official development assistance in recent years and uncertainty about its future size make it all the more imperative that the aid resources at hand be used more effectively and efficiently.

The fact that the substance of so much technical cooperation is better policy making and economic management is another reason for concern about effectiveness of technical cooperation. For better policy making and improved economic management are among Africa's highest priority needs. If the engine that is technical cooperation sputters, the result is slower long-term economic growth and delayed self-reliance.

The subject of this volume, then, is or should be at the heart of development policy. Many aid donors have long recognized its importance. But the United Nations Development Programme, as the world's principal supplier of technical assistance personnel and other technical cooperation inputs, has been most aware of its vital role and most sensitive to its shortcomings, as well as its successes. UNDP is concerned about whether these technical cooperation programmes are what the recipient countries need and want.

Soon after being appointed to UNDP as Assistant Administrator and Regional Director for Africa, in September 1983, I realized that, far from being at the heart of development policy, technical cooperation was a neglected factor. Most disturbing was that it was not managed seriously by the African governments because it is largely perceived, at best, as a free good and, at worst, as something imposed by the donors. I was astonished at the magnitude of these resources in the national context and became convinced that to ignore these issues would be a major error in resource-strapped Africa. I felt that the UNDP had a responsibility to bring technical cooperation back to the center of the development agenda, and to help African governments take an objective look at technical cooperation in light of their mounting frustration with the programme.

This is why, in the mid-1980s, we introduced the idea of NaTCAPs—National Technical Cooperation Assessment and Programmes—for use by countries to look systematically at the way technical cooperation is used for national development and to build their own capacity for better managing these resources. The majority of governments in Africa (34 to date) have started this exercise in evaluating technical cooperation, and through the NaTCAP process, UNDP has participated with these countries in these analyses and debates.

Through this process, we began to see that the same fundamental issues faced countries across the continent. Africa today is dramatically different from what it was at independence, and fresh ideas and new approaches must be tried out to meet the challenge of capacity building in the years ahead. We need to introduce reforms in the way we provide technical assistance. This is the reason for this book. We must rethink what technical cooperation is and what it is for, weigh the evidence, and propose new approaches.

Technical cooperation is a politically charged subject. This book presents a technical analysis of issues and proposes operational measures for reform, shying away from the ideological dimensions of the subject. Much other writing on this subject has been about ways to improve the impact of individual technical cooperation projects; this book focuses on technical cooperation as a major macroeconomic resource and explores the way in which it is allocated. Resources are normally allocated through budgeting and programming by those who control the process. Technical cooperation is programmed by donors, and not by recipients. Market forces do not apply. How, then, can we ensure that technical cooperation resources are allocated efficiently in the recipient country context?

This book was a joint undertaking between Elliot Berg and his collaborators at Development Alternatives, Inc. (DAI) and the staff of UNDP's Regional Bureau for Africa. In UNDP, Jacques Loup, Sakiko Fukuda-Parr, and Carlos Lopes, who spearheaded the development of the NaTCAP methodology and its implementation, were the primary persons involved. The book reflects the findings and views of these teams. It is not a policy statement of UNDP and does not reflect the official views of the institution.

Pierre-Claver Damiba
December 1992

ACKNOWLEDGMENTS

Many people contributed to this book, directly and indirectly. The staff of the Regional Bureau for Africa of the United Nations Development Programme produced, in the midst of their operational duties, a series of papers focusing on different issues of technical assistance in Africa. These papers, which are listed at the end of the bibliographic section, touched on many of the issues and problems that are the subject of this book and indeed are the source of much of its orientation. These papers provided much text and all of the statistical annex, and helped fix the guidelines for the organization of the volume. In addition, many of the authors of these background papers reviewed early drafts of this manuscript.

Several members of the Regional Bureau were associated particularly closely with the preparation of the book. Pierre-Claver Damiba, Director of the Regional Bureau for Africa between 1983 and 1991, played a special role. His awareness of the nature and extent of problems in technical cooperation in Africa, his insights about what had to be done to deal with them, and his sensitivity to the need for rethinking existing approaches brought the idea of producing this book to life. During much of its writing, he gave inspiration, encouragement, and ideas.

Staff members and consultants of the Bureau participated broadly in the preparation of the book, as did some UNDP field staff. Many wrote papers and memoranda. Many also provided useful comments, and the authors benefitted from the diversity of perspectives they provided. Bengt Sandberg, a consultant, participated in many brainstorming sessions and prepared several background papers.

Among staff members of the Bureau, Jacques Loup and Carlos Lopes gave particularly close reading to the manuscript and made numerous improvements. The statistical tables are the product of painstaking research by Marc de Bernis, with help from Alexis de Rocquefeuille. Sakiko Fukuda-Parr had the major responsibility on the Bureau side for putting the book together. She brought to the process her extensive knowledge of Africa and her long and diverse experience in matters related to technical cooperation. She was involved deeply in the writing and review process. Much of the volume reflects her insights and her ability to extract, express, and interpret the accumulated technical cooperation experience of the Regional Bureau for Africa as an institution.

Many staff members of Development Alternatives, Inc., the partner of UNDP in this venture, shaped the final product. Craig Olson was a major contributor. He provided drafts of several chapters and ideas for others. He was the main participant in the early phase of writing, and a perceptive reviewer of drafts. Channing Arndt, Ali Kemali, and Molly Phee helped in the research, under the continuing leadership of Graeme Hunter. The Office of Publications of Development Alternatives, Inc. prepared the text for publication.

The views expressed in this book are those of the authors and do not represent official positions of the UNDP.

Elliot Berg
December 1992

1

TECHNICAL COOPERATION UNDER FIRE

Growth in agricultural productivity is essential to economic development in Africa, as in most low-income regions. Yet progress has been slow despite serious efforts on the part of African governments and their external partners. Hundreds of millions of dollars have been spent on agricultural education and research, on agricultural extension, and on rural development projects of all kinds. Production nonetheless has barely kept up with population growth in most countries of the region and has lagged behind it in many.

Many factors, internal and external, explain this slow agricultural development including wars, civil strife, and political instability; fragile tropical soils and low or uncertain rainfall; macroeconomic policies uncongenial to agriculture; low world prices for agricultural exports; rarity of usable new technologies emerging from research stations; weak agricultural institutions; and excessively interventionist governments. One general factor has to be added to this list: disappointing results from the multiple technical cooperation efforts that mark recent decades.

Forty years ago, when formal technical cooperation came onto the world scene, there were high hopes that this form of aid, along with some capital assistance, would bring fundamental change, speeding

African agricultural growth and reducing rural poverty. But technical cooperation has proved to be a more demanding instrument, a less sure solution than its early sponsors imagined.[1] Why this is so and what to do about it are tough questions and the subject of much debate; they are what this book is about. But at least part of an answer to the question of what went wrong is at hand in a small story about the effort to improve agricultural productivity in one country—Somalia. The story predates the present misfortunes of that country; it illustrates common tendencies throughout Sub-Saharan Africa.

Like many African countries, Somalia received a great deal of external assistance in agricultural research and extension. Italy was active there before independence. Between 1964 and 1969, the United States Agency for International Development (A.I.D.), through the University of Wyoming, established and helped operate important research stations. In 1976, the United Nations Development Programme (UNDP) came in, contracting to a group of U.S. agricultural universities for many years, then to the Food and Agriculture Organization (FAO). By the mid-1980s, Somalia had benefitted from more than 20 years of relatively intensive technical assistance in agricultural research and extension.

Here is how a UNDP/World Bank assessment in 1986 summarized this long experience.[2]

> External assistance has undoubtedly provided considerable resources, both financial and human, to the areas of extension and research. At the same time, however, there have been some major shortcomings. First, there have been severe problems of coordination not only among projects and between projects and the permanent line departments but also between donors . . . for example, differences of view between the World Bank and USAID [U.S. Agency for International Development] over the implementation of the 'training and visit' extension system, . . . which has caused confusion in the efforts to design a national extension system. . . . The fact that UNDP/FAO provide support for research while extension services are assisted by USAID and the

World Bank has contributed to difficulties in linking research and extension. There have been divergent views, for example, with regard to appropriate programs for farming system research and on-farm trials.

Secondly, the discontinuity and fragmentation of assistance to agricultural research has meant that there has been little possibility of pursuing a consistent programme of investigation over an extended period of time. In fact, there is evidence that information collected and research results obtained in one phase of the assistance have been lost, or at least not been available to those working on other or later phases. As a result, the efforts to develop improved technical packages for farmers have evidently been disrupted. . . .

Thirdly, handling technical assistance through the (autonomous) Project Management Units responsible for individual investment projects has had a negative impact on the development of an integrated research and extension system. . . . Technical assistance in the agriculture sector, as it is currently provided, has (neglected) institutional development and (resulted) instead in a weakening of the permanent institutional structure. . . . The consequences of this are seen not only in the continued weakness of national institutions and reliance on technical assistance as well as perpetuation of projects over long periods of time, but more seriously contributing [*sic*] to the poor performance in the agricultural sector as a whole.

This analysis does not consider all the obstacles that have prevented effective transfer of technology and institutional capacity to African agriculture. But the analysis carries a clear message—that technical assistance has not brought to Africa the results expected of it. Experience with agricultural technical assistance has been happier in some other African countries, but the downside elements of the story are certainly familiar throughout the continent. The prevalence of these stories is one reason for the disenchantment with technical cooperation that has spread over the past decade, and especially over the past five years. There is a growing sense that technical cooperation does not work well, that as presently practiced it is ineffective, that such benefits as it brings are extremely costly, and that in any case it has little lasting impact.[3]

The current dissatisfaction with technical cooperation stands in sharp contrast to past attitudes. In the 1950s and 1960s, technical cooperation was seen, together with capital assistance, as the indispensable instrument in the struggle for world development. At present the feeling is widespread that it is more often misused than well used and frequently counterproductive.

RECIPIENT COUNTRY VIEWPOINT

Much more is now known about the nature and performance of technical assistance at the country level than was the case even a decade ago, most particularly as perceived by the recipient countries. A principal reason for this growth in information is the introduction by UNDP in the mid-1980s of the NaTCAP (National Technical Cooperation and Assistance Programme) idea—the notion that governments should do national technical cooperation assessments that describe the number and types of technical assistance personnel in the country, outline existing policies relevant to technical cooperation, and pinpoint problems, and that they should introduce better technical cooperation programming. The NaTCAPs have been financed mainly by UNDP, but implementation responsibility is in national hands. The products of the NaTCAP process include the recipient government's Technical Cooperation Policy Framework Papers (TCPFPs); a database on technical cooperation; and a system of rolling, multiyear technical cooperation programmes similar to public investment programmes. NaTCAP programmes have, to date, been launched in 30 African countries as well as in a handful of countries outside that region. About 20 of these governments have elaborated a national policy statement (a TCPFP) on how they wish to use technical cooperation resources in the future on the basis of the NaTCAP assessments.

The NaTCAP documents and related seminar proceedings present the most extensive and convincing evidence that exists about how recipients view the technical cooperation process, its contributions and shortcomings, and what needs to be done to make it more effective.⁴ The TCPFPs are prepared in most cases by national working groups led by senior officials, and are widely discussed within the government and with the donor community before being adopted as official government policy statements.

The technical cooperation policy statements reveal a number of recurrent findings with respect to technical cooperation effectiveness and commonly held views on the reorientation desired by recipient governments. The documents reaffirm the important role of technical cooperation in the development of the country. They underline that the primary objective of technical cooperation is to build national capacity. They acknowledge the past and current contributions of this aid and state that various kinds of technical cooperation will be needed for some time.

At the same time, almost all the national policy statements indicate that results of technical cooperation have been less than expected in building sustainable national capacity. For example, the government of Gambia concludes in its TCPFP that the "achievement of the twin objectives of skills transfer and institution building has generally been modest."⁵ The TCPFP of the Government of Benin points out that technical cooperation has not been effective in transfer of knowledge, and too much emphasis has been put on foreign experts, some of whom were of dubious quality.⁶ The TCPFP for Burkina Faso concludes: "The effectiveness of technical cooperation in terms of transfer of knowledge and reinforcement of national capacity is still limited."⁷

A recurrent theme in the national assessments is the concern about the persistent reliance on expatriate technical assistance personnel in spite of major efforts made in training. Either national training programmes and policies have been defective and have not given priority to appropriate areas, or donors have been perceived as requiring expatriate technical assistance personnel even when suitably qualified nationals are available.⁸ Whichever the case, virtually all countries express the view that technical cooperation is out of synch with the availability of trained national human resources. There is something wrong; the reliance on expatriate personnel does not diminish in spite of huge efforts and of gains in training of nationals. Even worse, unemployment among qualified local professionals is growing. Technical cooperation was justified by shortages in national human resources; it was expected to lead to self-reliance. Now, after decades of independence and concerted efforts at training, technical cooperation persists.

Although most documents recognize the important contribution of past technical cooperation, critical comments and calls for improved effectiveness abound. For example, the Burkina Faso orientation paper states: "Project documents which explicitly mention objectives of training and transfer of knowledge are the exception."[9] A survey carried out for the preparation of the Chad TCPFP reported the following judgments by Chadian project managers of the contribution of technical assistance to the success of various projects: important, 33 percent; mediocre, 42 percent; and weak, 25 percent.[10]

Governments are thus looking at the need for expatriate technical assistance personnel with greater skepticism. The policy statements of African governments state consistently that the aim is to keep the number of expatriate personnel to a minimum, and to promote the use of more highly specialized short-term consultants and of national consultants and experts.

Some countries have studied the numbers of expatriate personnel and compared them with the availability of trained nationals. Situations evidently vary from country to country and from one sector to another. There is little doubt that, in some countries, there are well-acknowledged shortages of qualified nationals, most typically in areas such as science teaching at the higher levels and medicine. In the Central African Republic, in 1985 there were 42 foreign doctors present but only 6 Central Africans graduated in this field.[11] In Mali, however, the situation was the reverse. In 1988, 152 qualified Malian medical doctors, 40 pharmacists, and 198 midwives were unemployed. In that same year, 73 foreign doctors were provided under technical cooperation schemes.[12]

That expatriate personnel are used even when similarly qualified nationals would be available is widely acknowledged in Uganda, where poor employment conditions in the civil service keep out qualified nationals. The technical cooperation surveys carried out in Burundi found that 34 percent of technical assistance personnel in the country were not university graduates. In areas such as agriculture, these personnel could have been replaced by nationals because graduates from the agriculture faculty were beginning to face employment constraints.[13] Burkina Faso's TCPFP and diagnostic studies report that, in 1989, 4,039 university graduates were employed in the Burkinabe civil service, 810 university graduates were unemployed (up from 485 in 1986), and 944 additional Burkinabe would graduate from the University of Burkina Faso in 1989.[14] At the same time, the number of expatriate technical assistance personnel with university degrees in the country was estimated at 800, roughly equivalent—the studies underscore—to the number of unemployed with university degrees. In some fields, expatriate personnel were present despite the availability of trained Burkinabe, whereas in other fields, such as health and certain teaching specialties (mathematics and physical chemistry), it was reported that there were not enough trained nationals to fill existing posts.

The principal explanation offered for the apparently anomalous employment situation in these countries is that budgetary constraints prevent the government from replacing expatriate personnel with unemployed nationals. The documents note that technical assistance personnel, whose services are provided free or nearly free, are used to supplement the state's operating budget. They also say that the services of expatriate personnel are often perceived as imposed by donors and are sometimes accepted by the state only to obtain the equipment and financial assistance that come with them. These observations are repeated in numerous other papers from many other countries. Often, although the finance or planning ministry may wish expatriate technical assistance personnel to be kept to a minimum, the line agencies clamor for more because they need extra hands, and argue that they should be given civil service posts if not technical assistance personnel.

The national policy statements do not call for a reduction in the overall level of but a change in its composition and its delivery mechanisms. Most of the statements recommend a reduction in long-term expatriate advisers in favor of more highly specialized, short-term consultants, national experts, and south-south cooperation, as well as training and equipment. According to a survey carried out in Chad, for example, 85 percent of high-level government officials considered technical cooperation indispensable, but only 23 percent considered the personnel component to be so.[15]

To the extent that technical assistance personnel are perceived as still needed, most of the diagnostic studies and policy statements argue in favor of cutting back on operational (substitution or gap-filling) personnel and expanding technical cooperation for capacity building or institutional development.[16] Several papers also point out that a major constraint to effectiveness lies with the personnel management policies and practices of the government; high turnover of staff and poor pay and working conditions do not make for well-motivated national staff.[17]

Another key area of concern consistently raised is that the recipient government should have a greater role in the selection and recruitment of technical assistance personnel as well as in their performance evaluation. Most policy statements call on donors to provide at least two candidates for posts, giving governments at least some choice in the matter.

Virtually all the national policy papers argue that the recipient government must get its own house in order to take a firmer hand in setting priorities and managing technical cooperation. These papers underscore the surprisingly large magnitude of resources tied up in technical cooperation activities. They point out that this form of aid cannot continue to be used as if it were a free good and that these resources, like any others, need to be put to best use. A recurring theme is that technical cooperation projects are poorly planned, programmed, and coordinated, largely because of a vacuum created by the absence of a strong central planning unit in government; the result is a multiplicity of uncoordinated, donor-driven technical cooperation projects. The Ghana TCPFP, for example, laments the "ad hoc project-oriented approach on the part of both donor and implementing agencies." This contributes to the fact that "the inter-relationships between recipient agencies—even those under the same umbrella organization—are often overlooked, thus seriously reducing the effectiveness of technical assistance."[18] The Uganda TCPFP points out that technical cooperation projects are not included in the rolling plan and therefore not in the budget, and thus not subjected to any planning process of the government. [19]

Thus most of the TCPFPs identify, as a major objective, the need for improved mechanisms to manage technical cooperation within government, including a process for priority-setting and implementation. Other points are repeatedly underscored in most of the documentation.

- Data on the number and type of technical cooperation projects actually operating in the country are thin, and few countries have usable national manpower inventories or assessments of need.
- The costs of expatriate technical assistants are extraordinary, both at the individual level and in the aggregate. The most humble technical assistance personnel earn easily 10 times the salary of government ministers, and the aggregate payments for these personnel are often a significant share of total public sector salary payments.
- Discrepancies in salaries and conditions of work, as well as cultural factors, often strain relations between the expatriate technical assistance personnel and national counterpart personnel, and can get in the way of achieving transfer of knowledge.
- The direct cost of technical cooperation, including personnel, can be more significant than is normally assumed and should not be ignored. These costs include contributions to salary payments and housing. Most governments avoid loan financing of technical cooperation as much as possible.
- Decisions on technical cooperation are usually taken without reference to budget priorities and without formal links to the public expenditure process.
- Suitably trained counterparts for expatriate personnel are lacking for many reasons: general skill shortages, fiscal pressure, low pay and poor working conditions, frequent

turnover of government staff, and departure of some of the most qualified staff to the private sector or to employment abroad.[20]

- Terms of reference for technical assistance personnel are poorly prepared, training plans are rare, and the "tied" nature of many donor technical cooperation programs limit the ability of recipient agencies to choose the personnel they want.

The debate on technical cooperation issues within African countries consistently reveals a strong interest in changing the mix of this form of aid. Most local officials involved say that more technical cooperation may be needed but not in the same form. They deplore the fact that the basic principles of the current system have not evolved since independence some 30 years ago. There is a clear fatigue with continued reliance on resident expatriate personnel. Technical cooperation seems off the mark in addressing priority issues, which include what to do about unemployment among trained professionals and how to make national institutions function better. The national policy statements on technical cooperation unequivocally confirm that the recipient countries want to play a more effective role in determining the use of technical assistance.

DONOR CRITICISMS

For a long time, technical assistance was rarely criticized in donor circles; it was regarded as a good thing, the indispensable vehicle for technology transfer. Beginning in the 1970s, however, dissonant voices appeared. At first, expressions of dissatisfaction were not sharply focused; they involved general grumbles about cost or effectiveness.[21] Sometimes analyses of technical cooperation—its purpose, its forms, its effectiveness—were included in more general assessments of development assistance.[22] But few observers concentrated their analyses uniquely on technical cooperation.

Over the last several years, however, focused assessments have increased in number and have become more critical. Several aid agencies have conducted reviews of the effectiveness of technical

CHAPTER 1 Technical Cooperation Under Fire 11

cooperation and numerous international conferences have been convened to discuss related issues.[23]

One result of the growing debate has been a convergence of opinion on the nature of technical cooperation problems. It would be going too far to say that a tight consensus exists among the critics. Observers have different views about the importance of various dimensions of the problem and about the need for reform. But the outline of a consensus critique is now clear; it embraces, to a great extent, the criticisms expressed by recipients.

Some recent donor judgments are indeed more explicit and more sharply critical than those of recipients. A 1989 assessment by UNDP staff presents a comprehensive list of deficiencies.[24]

> A major part of technical cooperation expertise is allocated to filling operating positions . . . and not giving adequate attention to the longer term tasks of building African institutional and individual capacities. Counterpart development is not working. Training activities are slighted. The demands on expatriate operating personnel, such as teachers, medical staff, managers, ministry technical staff, etc., to get the work done divert attention from the longer term tasks of developing individual and institutional capabilities.
>
> A second concern is one of quality in technical cooperation personnel. Technical cooperation technicians are used for a variety of tasks for which they may not be equipped . . . mixing policy, institution building, and technical and operating responsibilities. . . . Often specific work objectives are not spelled out. Problems of quality (of technical assistance personnel) are also evident in . . . (their) lack of knowledge of local conditions and insensitivity to local cultures.
>
> African countries have well educated and competent professionals who are not being used effectively in their country's development programs. African governments and donors are at times too quick to bring in outside expertise without exploring the African capabilities available at home or that could be attracted to return.

Technical cooperation project objectives are not well defined, too diffuse, overambitious and not linked to government programming and budget processes.

The definition of technical cooperation projects and technical services is supply driven and determined by donors rather than guided by well defined priority demands.

The volume of technical cooperation, the number and diversity of projects and donor procedures, is overwhelming African government abilities to coordinate and manage. Efforts to coordinate aid in general are sporadic and of uneven efficacy.

Evaluations of infrastructure projects when focused on the question of sustainability have found that insufficient attention had been given to building sustainable institutions to operate and maintain these investments.

A comprehensive World Bank review of internal studies, project documents, evaluation reports, and other sources concerned with technical cooperation for institution building summarized its assessment of past efforts as follows.[25]

> Various internal reviews carried out in recent years, mainly by the Operations and Evaluation Department (OED) and the Africa Region, have pointed to serious weaknesses in the way in which technical assistance, particularly ID [institutional development]-related technical assistance, is managed by the Bank. The frequently hasty and poor design of technical assistance projects, in part attributable to inadequate diagnosis of technical assistance needs, tends to be compounded by defects in implementation such as recruitment delays and difficulties in finding suitable consultants (particularly for training), problems associated with the employment of long-term expatriate advisers, lack of adequate counterparts, lax supervision by the Bank, poor coordination with other donors, and the inadequate administrative capacity of the borrower. These are especially serious problems in Africa, as well as in a few least developed countries elsewhere (e.g., Bangladesh); it is in these parts of the world that

the phenomenon of 'supply-driven TA' [technical assistance] is most commonly encountered.

These critical assessments are typical of recent writing and are echoed in the conclusions of many recent conferences on technical cooperation. After noting the "strong feeling of unease" and growing doubts that the results of technical cooperation are commensurate with its high costs, the authors of one report observe:

> Some critics do not hesitate to compare technical cooperation with the 'Marxian state': while it was designed to 'wither away' it became in reality an ever-increasing monster. At the Maastricht Conference 'Beyond Adjustment' (July 1990), the Chairman of the DAC [Development Assistance Committee] argued that 'it was wasteful and unacceptable that technical assistance was taking jobs away from able local people.' The delegate of Mauritius paraphrased Oliver Cromwell to reflect a widely shared African view on technical cooperation: 'You have stayed here too long for all the good you have been doing. In the name of heaven, go!'[26]

In an article in the *Journal of the Society for International Development* titled "A Future for UN Aid and Technical Assistance," a high-ranking official of the United Nations Childrens Fund (UNICEF), expressing his personal views, comes out not far from the Cromwellian paraphrase:[27]

> I believe that the vast bulk of technical experts and expertise at present provided by the UN and donor system has outlived their usefulness . . . judged by the criteria for which they have been provided: the provision of specific technical expertise or experience which is not available among nationals of the country . . . for a limited period until national personnel have acquired the training and expertise to take over the job. . . . (Far) from diminishing, the numbers of technical experts provided has grown decade-by-decade since the 1950s. . . .

> [Costs have] reached extraordinary disproportions. In Tanzania, for example, the total cost of technical assistance in 1988 was some $300 million, of which at least $200 million represented the salaries, per diems, housing allowances, air travel and other direct costs of the 1,000 or so international experts provided as the core of technical assistance. In contrast, the total salary cost of the whole civil service in Tanzania in the same year, including administrators, clerical staff, teachers and health workers, was $100 million. The situation in Tanzania is not untypical. . . . The time has come to rethink the purpose of aid and technical assistance within the UN system.

Very few observers see this form of aid as so total a failure. Many point out the dubious logic of measuring the effectiveness of technical cooperation only by whether it has increased or dwindled. One can argue that the modern institutional landscape in Sub-Saharan Africa—from science faculties to television systems—is prima facie evidence of the success of technical cooperation.

The past 30 years, for all their disappointments, have witnessed extraordinarily dense institutional growth in Sub-Saharan Africa, and growth in capacity to manage: central banks are now locally run, as are new school systems, agricultural research stations, power plants, airlines, armies, and universities. It is not atypical that the University of Malawi, staffed entirely by expatriates in the early 1960s, is now staffed 90 percent by Malawians and that the number of British experts in the central government of that country fell from 800 in 1980 to fewer than 200 a decade later. That many of these institutions do not work very well is more a function of their recent creation, their rapid growth, and—most important perhaps—the lack of an adequate enabling environment in the public sector than of any intrinsic failure of technical cooperation.[28]

Moreover, almost no evaluation levels an unqualified attack on all technical cooperation. The consensus judgment is that aid for operational support—so-called gap-filling or substitution technical cooperation—has worked well.

The problem is that technical cooperation has left behind too little indigenous capacity. It has not worked itself out of Africa and not built self-reliant institutions as was originally foreseen. This is where the perception of ineffectiveness comes from. The perception is based mainly on the balance sheet of technical cooperation after three decades of African independence. Real expenditure in this area has not much decreased, tens of thousands of expatriate personnel are still present, and many institutions critical to the functioning of a modern state simply do not work very well. It would be wrong, however, to heap on technical cooperation alone the blame for this state of affairs, just as it would be wrong to attribute to it alone the positive institutional developments of the past decades.

There is, in any event, strong evidence of the failure of institution-building technical cooperation: most recent formal evaluations indicate a very modest level of effectiveness.

FORMAL EVALUATIONS

Formal evaluations of the technical cooperation process are few in number, so it is worth giving those that exist special attention. Because evaluation of this form of aid presents some particular methodological problems, we review briefly some of the technical issues before setting out the results of the more significant evaluations.

Problems of Methodology

Four problems of methodology stand out: unweighted multiple objectives, unclear and intangible measures of output or effectiveness, use of proxy indicators, and multiple determinants of performance.

Multiple Objectives

Effectiveness has to be measured against objectives. But few technical cooperation projects have single and precise objectives. Even so-called engineering types of technical cooperation (designing a road, computerizing a debt management system) usually involve on-the-job training, and most nowadays include formal institution-building components. Obvious problems of weighting arise. How can one measure the effectiveness of economists assigned to a planning ministry to produce a manual on project analysis, assist in the drafting of a development plan, train ministry staff, and assist in plan implementation? How much weight should be given to the quality of the documents produced, the training done, the planning procedures the economists introduced, or the bad projects turned back because of the economists' presence?

Intangible Output Measures

How do you measure the impact of an energy specialist who is asked to write an energy sector program? Should it be by the quality of the document produced? Obvious issues of subjectivity arise. Or is the appropriate measure the difference in quality between the document of the specialist and one that might have been produced by local staff? But subjectivity issues loom even larger here.

How does one measure the extent to which institutional capacity has been strengthened? There has been very little theoretical work on indicators of institutional development and even less empirical work. The state of the art is summarized in a recent World Bank report:

> Effectiveness and especially impact of ID-related technical assistance are difficult to measure. The nature of institutional development makes it difficult to specify measurable outputs and to quantify achievement indicators. This is especially true for relatively new areas such as public sector management, for which the conceptual and methodological bases have not yet been agreed upon.[29]

Use of Proxy Indicators

If the objectives of a technical cooperation project are well defined (which is not always the case), one can usually identify the direct, immediate outputs expected. A training programme can count the number of people trained. An agricultural research programme can be evaluated, at least partially, on the quantity and quality of research papers produced, or on rates of adoption of new seeds or methods. An institutional-strengthening project might use such output indicators as in-service trainees trained, procedure manuals produced, and computerization attained.

It is much more difficult, however, to derive indicators of the long-term impact of technical cooperation programs on development. The development impact of trained people, for example, can be measured only in terms of their increased productivity over time. The ultimate indicator of the development impact of agricultural research is its contribution to increased production. The success of an organization-strengthening programme can be measured only by assessing the efficiency and viability of the institution over several years.

Almost all technical cooperation evaluations (other than those based entirely on perceptions) are therefore derived from proximate or intermediate indicators that are based on immediate outputs, not ultimate indicators of effectiveness. These output indicators are sometimes used as proxies for ultimate indicators. Ultimate impact is forecast on the basis of projections about the cumulative effects of the immediate outputs. It is assumed, for example, that procedure manuals or computerization will lead to greater efficiency in a beneficiary organization.

Actually, most evaluations of effectiveness make little use even of proxy indicators. They rely instead on the judgments of key actors and especially on the perceptions and judgment of the national staff of beneficiary organizations. Some critics tend to dismiss this type of evaluation as unscientific. Others applaud it.[30] In any case, there is little else at hand.

Multiple Determinants of Performance

Even if technical cooperation evaluations found usable ultimate indicators, they would have to disentangle the effects of this aid from those of other variables that contribute to organizational effectiveness.

If a commodity marketing organization that has received technical assistance increases its sales and holds the line on costs, what share of the improved performance is attributable to the technical assistance and what share to, say, an improvement in commodity prices, a change in management, or higher salaries for the staff of the organization? If the same organization fails to improve its performance over a number of years, to what extent should poor technical assistance bear the blame, rather than low commodity prices, management turnover, or deteriorating staff salary incentives? In the latter case, it can always be argued that the technical assistance prevented a bad organization from getting worse.

Evaluation Record

Notwithstanding these methodological difficulties, several substantial efforts at systematic evaluation have been undertaken since the mid-1980s, in addition to the numerous diagnostic studies done under NaTCAP auspices and broad summary reviews. The following sections review the main findings and conclusions from some of these major evaluations.

The Cassen Report

Of all the evaluations done in the past 10 years, the Cassen report is the most positive about the effectiveness of technical cooperation. Chapter Six of the book *Does AID Work?* by Robert Cassen and Associates (1986) deals with this issue. Its author is Robert Muscat, who also published separately many of his findings.[31] The author argues that because engineering-type technical cooperation has had a "very high success record" he concentrates on "institutional and human capacity-building" technical cooperation because this type of technical cooperation has been most criticized.[32]

Overall, according to Muscat, technical cooperation has been a clear triumph: it has "played a major role" in the "vast creation of institutional and human capabilities throughout the Third World in the three-decade history of international development assistance." The flaws of technical cooperation are real, but are magnified when the record is disaggregated by sector, type of assistance, donor, and other variables.

Muscat's review of published evaluations leads him to conclude that the performance of between one-half and two-thirds of all technical cooperation projects for institutional development were judged satisfactory, with one-third judged fully satisfactory. Only 10 to 15 percent were judged to be outright failures. Even these generally good results, he believes, understate effectiveness because the literature tends to concentrate on "soft" technical cooperation, where effectiveness is more difficult to measure. He concludes: "The record also suggests that critics who are inclined to make a blanket condemnation of technical cooperation projects have lacked discrimination, drawing strong conclusions from selective evidence."

The Muscat chapter in the Cassen report surveys performance by sector and reaches the following conclusions.

- Among the most successful technical cooperation projects have been those aimed at strengthening central banks or improving the process of financial intermediation.
- For meteorology, "there is prima facie evidence of widespread impact of technical cooperation," which was absolutely necessary for the development of the meteorological agencies in most developing countries.
- For forestry, there have been positive proximate effects, but the jury is still out with respect to the ultimate effects.
- For agricultural research, the most remarkable positive achievement of technical cooperation concerns the development of new grain varieties associated with the Green Revolution. A review of 39 evaluations of A.I.D. agricultural research projects found that the performance of one-half of the projects was satisfactory or better and one-half were less than satisfactory. A review of 92 UNDP/FAO technical cooperation projects found that "of 33 agricultural research institutions operating for at least four years, two-thirds of the institutions were active, ongoing facilities, and the great majority of trainees had returned to their institutions." Yet many agricultural research

institutions had "deteriorated" after technical cooperation was withdrawn; the main reason cited for these institutions' lack of self-reliance was that the technical assistance had concentrated too much on the actual agricultural research and not enough on training and institutional development.

- Technical cooperation for pastoralism has been regarded as a failure.
- With respect to education, in the 1960s and 1970s, as well as during the colonial era, technical cooperation has played a major role in the creation and expansion of educational institutions and in the formal education and training, both in-country and overseas, of large numbers of Africans. More recently, education projects have concentrated on such areas as educational planning and reform of rural education while continuing with the provision of fellowships and other types of participant training. Muscat cites a review of about 20 evaluations of education projects from DAC evaluation correspondents that found that almost all of the projects were successful within the limits of the objectives and available measures (for example, enrollment). A review of 55 World Bank education projects found that most components were successful, an important exception being educational planning. A review of 25 UNDP/UNESCO (United Nations Economic, Social and Cultural Organization) evaluations aimed at innovation and reform in national educational systems found that, in terms of proximate effects, two-thirds were successful, 13 percent were unsuccessful, and the remainder had mixed results. In terms of ultimate effects, however, only one-third were rated good, the rest as mixed or deficient.
- In health, the most striking successes of technical cooperation have been in "functional" or "campaign" approaches, such as the complete elimination of smallpox or the development of oral rehydration techniques to combat diarrhea in infants. In other health projects, effectiveness has been mixed. A review of 39 health projects contained in DAC effectiveness reports notes

that three quarters of the projects were considered wholly or partially effective.
- In population, the Cassen report underlines the difficulties of measuring the effectiveness of technical cooperation but ventures, nevertheless, that it has had a "substantial impact on population policy, a marked but varying impact on family planning programs, and only an indirect effect on fertility itself."

Muscat's assessment of technical cooperation effectiveness is so much more favorable than other recent evaluations that it demands some explanation. Part of the reason may be timing. He writes in the middle 1980s and looks at data from earlier years, before the wave of dissatisfaction approached its peak and many negative written assessments appeared. His assessment is based entirely on secondary materials. Also, he tends to be skeptical about negative assessments. Evaluations are said to have negative bias; they concentrate on problems and weaknesses. He dismisses many criticisms of technical cooperation on the grounds that they are subjective and based on selective evidence. As a result, Muscat inadequately addresses the process-related issues that are so central to many of the recent evaluations and their negative findings on the capacity-building effects of technical cooperation.

Muscat also concentrates on what technical cooperation programmes achieved rather than questioning whether they were addressing the most important objectives. His suggestions for improvement in Africa do not indicate an awareness of deep-seated problems: he calls for more time, education, and use of nongovernmental organizations and of noneconomists in project design. These recommendations give no more than a hint that the author recognizes the systemic problems of technical cooperation as a capacity-building instrument.

The Forss Report

An evaluation of an entirely different tone and texture—and one that is among the most widely cited in recent years—is a report by

Kim Forss and four colleagues entitled "The Effectiveness of Technical Assistance Personnel," which was published in 1990.[33] The Forss report consists mainly of case studies of 55 projects in Kenya, Tanzania, and Zambia that were financed by the Nordic countries (Denmark, Finland, Norway, and Sweden) and that employed technical assistance personnel from those countries. The study is supplemented by analyses of the national manpower development policies of the three African countries.

The methodology consists of a review of project documents and interviews with a sample of 324 technical assistance personnel; local personnel, including technical assistance personnel counterparts and members of beneficiary groups; and aid agency officials. A questionnaire was also mailed to 365 persons who had been employed as technical assistance personnel on the 55 projects, although this questionnaire generated only a 37 percent response rate.

The report found that, despite growth in the pools of trained local manpower, demand for technical assistance persisted at a relatively high level in the three countries. Local manpower was found to be sufficient to cover demand in such areas as general administration and accounting, and technicians at the certificate and diploma levels were in adequate supply except in certain specialized skills in manufacturing. Shortages of skilled and experienced managers existed in all three countries.

The report noted, however, that the effective demand for technical assistance personnel was not always directly related to the availability, or lack thereof, of local manpower. Technical assistance personnel were seen by ministries as a way to augment their personnel on an extrabudgetary basis. The demand for this personnel, particularly those functioning as project implementers or "controllers," was also created by the proliferation of projects; donors frequently created parallel project implementation units staffed with expatriates because they felt that they could not entrust project implementation and project funds to existing government agencies.[34]

Another reason for the continuing demand for technical assistance personnel, especially at the managerial level, was the inability of governments to attract or retain local personnel because of low salaries and unattractive working conditions; private sector opportunities were more attractive, especially in Kenya. Even when local manpower was available, public sector officials sometimes preferred technical assistance personnel because of their greater experience and a perception that they could "do the job faster."[35] The report found, nevertheless, that local manpower could fill 204 of the 324 technical assistance personnel positions "if problems of lack of funds, etc. are solved."[36]

In terms of effectiveness, the principal conclusion of the Forss report was that technical assistance personnel were usually highly effective in operational positions but much less effective in transferring skills and in contributing to institutional development.[37] The report found that technical assistants may even have had a negative effect on institutional development because they tended to create or enlarge institutions that are not sustainable. Overall, the report found that the impact of technical assistance personnel did "not stand in favorable relation to the cost."[38]

The attitude of host country personnel to expatriate personnel was generally unfavorable on many counts. Host country officials acknowledged the effectiveness of most technical assistants in producing outputs, but most often felt that there were far too many of them and that they were imposed on the host country by donors as a requirement for obtaining project assistance. Technical assistants were often envied or resented because of their "controller" functions; because of their privileged access to project facilities and equipment, especially vehicles; and because of the large disparities in their salaries compared with those of local officials.[39]

One of the principal reasons that technical cooperation was found to be ineffective in skills transfer was the breakdown of the expert-counterpart relationship. Only about half of the technical assistance personnel who were supposed to have counterparts actually had them. Local employees generally found counterpart positions unattractive because of their perceived lack of importance and power and because of ambiguity about where such positions fit into the career ladder. Often, therefore, these posts went to less capable local staff.

Expatriate technical advisers, in turn, were reported to have become rapidly frustrated at the poor quality or the lack of interest of the counterparts assigned to them. Faced with pressures to get on with the job and produce measurable outputs, these expatriate technicians found themselves devoting less time to the training of counterparts and more time to their operational responsibilities. In addition, the personnel in the sample were usually recruited on the basis of their technical skills and seldom if ever because of their skills as trainers or their cultural sensitivity.

World Bank Reports

The World Bank was originally created as a capital assistance institution while other multilateral institutions, notably UNDP, were established to provide technical assistance. In its first two or three decades, therefore, most of the Bank's lending was for large infrastructure projects and rural development. The Bank did provide some technical assistance in conjunction with these projects, but most of it was of the so-called "hard" variety—in other words, feasibility studies, engineering services, and project management.

In the last 20 years, however, the World Bank has devoted a much larger portion of its resources to technical assistance, especially with respect to Sub-Saharan Africa.[40] Most of this increase has been "soft"—that is, the provision of expert services and training aimed at capacity building and institutional development.[41] The main reason for this shift was that evaluations of World Bank investment projects had consistently shown that difficulties of project implementation and sustainability were attributable less to technical factors than to shortcomings in management and institutional performance.[42] Technical assistance was seen as the input best suited to resolving these management and institutional problems.[43]

As the World Bank's emphasis on technical assistance has increased, so too has the number of evaluations aimed at assessing its effectiveness.[44] The general thrust of the findings of the World Bank's evaluations is that capacity-building technical assistance usually has low success rates, much lower than hard or engineering-type technical assistance.

- A 1984 evaluation reviewed 176 project performance audit reports issued through December 1982 on completed projects in agriculture and transport sectors in Sub-Saharan Africa. One hundred and sixteen projects contained institutional-development measures. Only 5 percent of the projects were judged to have achieved their institutional-development objectives in full, while 17 percent failed entirely; the rest fulfilled immediate objectives "partly" or "mostly." But in more than 60 percent of the projects, the institutional-development component was judged to have had zero or slight impact on the target institution.[45]
- A 1985 study, based on field visits and examination of completion reports and audits, looked at International Labour Organization (ILO), UNESCO, and World Bank experience with capacity-building projects in education in five African countries (Cameroon, Madagascar, Mali, Sierra Leone, and Zambia). Little management capacity had resulted, mainly because of insufficient emphasis on training.[46]
- A 1989 review of 366 Sub-Saharan African projects with institutional-development components found substantial results in 22 percent of the cases, negligible results in 26 percent, and partial success in 52 percent. The "partial" category is believed to include many projects with very slight positive results.[47]
- The "1990 Annual Review of Evaluation Results" found that the performance of 62 percent of completed technical assistance projects was satisfactory. However, the success rate for the institutional-development components of all the projects evaluated was only 25 percent, with 42 percent assessed as having negligible results concerning institutional development. The failure rate was particularly high for technical assistance projects linked to structural adjustment, and higher for Africa than for other regions.[48]
- Studies undertaken in connection with the work of the 1991 Technical Assistance Review Task Force confirmed poor results for technical cooperation related to structural adjustment.

"Substantial success" was achieved in less than one-third of the projects examined.[49]

- Despite repeated calls for more careful design of technical cooperation projects for institutional development, a recent study concludes that most World Bank projects continue to give insufficient attention to institutional-development issues.[50]
- In 1989, the Operations Evaluation Department evaluated the effectiveness of 19 free-standing technical cooperation projects aimed at improving public sector management in seven African countries. Only five were found to be successful and only three to be wholly satisfactory. A later independent evaluation of one of these projects found its success rating to be unduly optimistic.[51]
- In 1991 the World Bank set up a Task Force to assess Bank performance in technical assistance lending. The Task Force received 74 responses to its questionnaire—about half of them from Africa.[52] The responses were not reassuring, especially those from Africa.
  — Only 47 percent of the African responses said that the technical assistance objectives were clearly defined and understood.
  — Only 39 percent felt that the process for selecting consultants was satisfactory.
  — Almost 70 percent of the respondents from all regions said they would not have used the same amount of technical cooperation if they had to pay for it out of local budget resources. The proportion of African respondents who gave this opinion is not broken out, but it was evidently much higher than 70 percent, because the Asian respondents were more positive.
  — Overall (all types of technical assistance and all regions), some 60 percent of respondents were strongly affirmative in response to the question of whether technical assistance project objectives had been met. But 28 percent in Africa (and 20 percent in the Latin American and Caribbean region) gave strongly negative responses.

- The Task Force also reviewed recent internal assessments of the Bank's past technical assistance. The results confirm the central theme of past evaluations: that technical assistance projects or components aimed at institutional development have been largely ineffective.
  — According to the Operations Evaluation Department's "1990 Annual Review of Evaluation Results," only one quarter of the more than 300 projects with institutional-development objectives had "substantial" results; another quarter were clear failures. Of the projects identified as technical assistance, an even higher share—42 percent—had negligible institutional-development impacts. This is true for all regions. The numbers for Africa alone are undoubtedly worse, but these are not broken out.
  — A sample of recent project completion reports reviewed by the Task Force reveals that, although more than three quarters of hard technical assistance projects for engineering or preinvestment purposes were considered successful, only a quarter of institutional-development projects (public sector management) were so rated.
  — The Task Force undertook a review of project completion reports and other evaluations for 45 investment projects with technical assistance components. The African sample included 12 projects from 10 countries and covered eight sectors. Worldwide, the impact of these technical assistance components was judged negligible to partial. In the African sample, six projects had virtually no results and six had partial results.[53]
  — The Task Force reviewed the 99 adjustment lending operations—structural adjustment loans (SALs) and sectoral loans (SECALS)—approved in the 1980s for which project completion reports were submitted. Fifty-four of the 99 projects were supported by 50 technical assistance loans, either as components or free-standing. Twenty-eight of these were in Sub-Saharan Africa. Only two of the 28 African projects recorded "substantial" results, though only

four were judged to be outright failures. Overall, substantial success was reported in fewer than one-third of the projects. The Task Force report comments that this "provides little grounds for satisfaction . . . ," and concludes: "The effectiveness of adjustment-related technical assistance is disappointing."

The report's annex on evaluation of technical assistance results concludes, after a review of the entire Bank project portfolio: "These data strongly support the preceding evidence of poor performance and the need for radical improvement in the Bank's management of technical assistance, particularly for institutional development."

Summing Up

Despite their patchy coverage and their occasionally rag-tag methodologies, the evaluations that have been conducted on technical cooperation effectiveness strongly confirm several principal conclusions that emerge from more casual observation.

The first is that blanket statements cannot be made about the effectiveness of all technical cooperation, which consists of diverse inputs (people, training, and equipment, for example) and many objectives. Its effectiveness varies according to type, economic sector, specific input, specific objectives, delivery mechanisms, and region of operation.

Second, most of the criticism concerns lack of positive impact on capacity building or institutional development. Institutional weaknesses are acknowledged to be a primary cause of persistently poor development performance in Africa, and technical cooperation—particularly technical assistance personnel—continues to be regarded as a primary vehicle for strengthening weak institutions. Both donors and recipients agree, however, that technical cooperation has been ineffective in achieving this objective. They therefore call for substantial reforms. Only the 1986 evaluation by Muscat fails to come down strongly on this point, although even his evaluation admits to weak performance in Sub-Saharan Africa.

Third, most of the criticisms of technical cooperation are leveled particularly at long-term expatriate technical assistants—the resident experts: whatever their ostensible purpose, most technical assistants are used not for capacity building but for support of operations; expatriate technicians are outrageously expensive; and they displace available local skilled personnel. Paradoxically, most observers agree that technical cooperation personnel are effective in their operational roles. But they are found wanting at transferring skills to indigenous counterparts (capacity building) or at creating systems that will sustain their work once they leave the country.

The training dimension of technical cooperation also comes in for general attack, on several grounds. With respect to formal training, the major criticism is that training programmes related to technical cooperation are ad hoc, and not based on systematic assessment of manpower needs. The informal training component of technical cooperation—on-the-job skill transfer to counterparts who work alongside resident experts—has proved highly imperfect in practice. The main reasons for this include poor selection, high turnover or simple unavailability of counterparts, and the preference of most resident experts to do the work themselves rather than train others to do it.[54]

This last factor is crucial and enters even when technical cooperation objectives and terms of reference are clearly specified. The terms of reference of a resident adviser may say to build capacity by training people and strengthening the agency. But the adviser will most often end up mainly providing operational support. This occurs because of the internal values of the expert; because the prospects for effective institutional development are so unpromising; or because it is what the expert's bosses want, can see, and will appreciate.

Fourth, there is wide agreement that of all types of technical cooperation, that aimed at institution building is the least effective because it presents the most profound challenges.

- It deals with issues that are often vaguely defined, almost always culture-bound, and frequently threatening to one or another set of stakeholders. Its outputs are not like those of engineering-type technical cooperation—roads, buildings, or power plants—or even like the reports and manuals that some technical assistance personnel leave behind; its outputs are usually much less easy to identify. For engineering-type technical cooperation, a high degree of consensus usually exists between the donor and the recipient on the need for and the intrinsic value of the output. The expected outputs of technical cooperation for institutional development, in contrast, are less tangible, harder to identify and measure, and may not be seen as equally desirable by both the donor and the recipient.

- Institutional development aims, by definition, at inducing changes in human and institutional behavior, and such changes may come into conflict with existing bureaucratic norms and values or may threaten organizational alignments and coalitions. One obvious example is an attempt to introduce a system of merit promotions based on performance evaluations; this approach is likely to conflict with the patron-client system prevailing in the organization.

- Expatriate technical assistants who introduce technocratic organizational reforms carry with them different cultural baggage than nationals, on everything from management styles, to attitudes toward authority, to how achievement-oriented and ascriptive factors ought to be weighted in hiring, firing, and promotion decisions. Expatriate advisers are almost always recruited on the basis of their technical skills rather than on the basis of interpersonal attributes such as empathy, cultural sensitivity, and adaptability. And, to top it off, terms of reference stress technical performance over capacity building. It is hardly surprising, given these factors and the delicate cultural paths that have to be navigated, that failure rates of technical cooperation for capacity building are high.

These problems of technical cooperation can for convenience be grouped into four categories. The first consists of what we call delivery system failures. These are of two types. Weaknesses in project identification, design, and implementation make up the first type. These weaknesses concern faulty nuts and bolts: hasty project selection, excessive complexity, inappropriate terms of reference for technical assistance personnel, poor supervision, and the like. The second type of delivery problem is more fundamental or systemic: the fact that the favored instrument of technical cooperation delivery—the resident expatriate expert along with local counterpart—has proved unsuitable for capacity building.

The second set of problems revolves around deficiencies in the management of technical cooperation. Weak coordination and management are primary factors contributing to technical cooperation ineffectiveness.⁵⁵ Responsibility for management should belong to recipient government agencies, but African planning agencies and other coordinating entities have not been able to manage their country's technical cooperation inflows. One reason is that the large volume and increasing diversity of technical cooperation resources have, in many countries, overwhelmed local management capabilities. Another reason is that donors often prefer to maintain tight control over the selection of projects and the management of resources.

As a result, the management of technical cooperation in Africa remains, to a large extent, in the hands of donors. The selection of projects is donor driven, and technical cooperation involves transactions between "unequal partners."⁵⁶ In some countries, donors have established management procedures that all but bypass local authorities. Psychologically, and in actual practice, many technical cooperation projects are perceived as donor projects, rather than government projects. The governments' weak ownership of these projects has spawned—understandably—weak local commitment and frequent indifference to project sustainability. Also, the large number of donors and donor projects, combined with local management weaknesses, has resulted in poor coordination of projects and programs among donors. Inefficiencies in allocation of technical cooperation resources, expressed in redundant or even contradictory projects, are therefore not rare.

The third problem has received little attention in the literature but is basic: the lack of an effective market for technical cooperation. One aspect of this situation is the tied nature of transactions: "buyers" usually have to take the whole package—experts, training, and equipment. More fundamental, technical cooperation is mostly grant funded. And even if it is not, the users (operating agencies) don't pay.

The market for technical cooperation is therefore largely a market without prices. On the demand side, the money costs of this form of aid to recipient agencies are minimal. With opportunity costs close to zero, user agencies have little reason to decline technical cooperation projects, to choose wisely among alternative projects, or to economize during project implementation. On the supply side, donors have many reasons for urging recipients to accept projects and often find themselves in competition with one another for access to the client. The incentive structure in the technical cooperation market is thus out of joint. It encourages proliferation and redundancy of projects and dilutes local commitment to their effective implementation.

The fourth and final set of problems relates to the work setting within which technical cooperation operates on the ground. In most African countries, civil services suffer from serious problems of morale and performance because of low pay and poor working conditions. The situation breeds low job commitment—high turnover, pursuit of private activities to make ends meet, limited interest in job-specific training. In these working environments, it is utopian to expect externally provided technical cooperation resources to make much of a difference in terms of capacity building.

These are the problems that have to be confronted and overcome if technical cooperation is to realize its full promise. They are addressed in the following chapters. Chapter Two provides background: it sorts out definitional issues, describes the historical setting of technical cooperation in Africa, and gives the statistical record. Chapter Three considers delivery system issues, Chapter Four deals with the management of technical cooperation, Chapter Five analyses the functioning of the technical cooperation market, and Chapter Six considers the problems of poor work environments. Chapter Seven presents conclusions and recommendations.

In each of these chapters, other than Chapter Two and Seven, the approach is the same. The problems are described and proposed reforms are assessed. The emphasis throughout is on the latter, on how technical cooperation can be made more effective and play more fully its vital role in the development of Africa.

NOTES

1. "Technical assistance" and "technical cooperation" are used interchangeably throughout this book, as they are in most discourse. Both are defined as a set of inputs (technical assistance personnel, training, and equipment) used to produce a specified set of outputs: enhanced skills and strengthened institutional capacity. Chapter Two contains an extensive discussion of concepts and definitions.

2. UNDP and World Bank, "Somalia: Report of a Joint Technical Cooperation Assessment Mission," 1985, pp. 17-18.

3. See Organization for Economic Co-operation and Development, Development Assistance Committee, "Principles for New Orientations in Technical Cooperation," Paris, 1991; and J. Bossuyt, G. Laporte, and F. van Hoek, "New Avenues for Technical Cooperation in Africa," European Center for Development Policy Management, Occasional Paper, Maastricht, 1992. A summary review is presented in S. Adei, "Overview of Technical Assistance/Cooperation," UNDP internal paper, October 1990.

4. For a brief summary and analysis of 18 TCPFPs, see Paul Geli, "Review of Technical Cooperation Policy Framework Papers (TCPFPs)," UNDP, October 1991.

5. Government of The Gambia, "Technical Cooperation Policy Framework Paper," 1990, p. 11.

6. "La Coopération Technique n'a pas effectivement joué le rôle qui aurait dû être le sien vis-à-vis des cadres nationaux ainsi que dans le développement des institutions." République populaire du Bénin, "Document de Politique de Coopération Technique," mai 1989, p. 9.

7. Burkina Faso, Ministry of Planning and cooperation, "First Round Table Conference for Burkina Faso," May 1991, p. 54.

8. For example, see République du Mali, "Document d'Orientation sur la Politique de Coopération Technique du Mali," 1991, which states: "Une partie de l'assistance technique occupe encore des postes d'exécution et de substitution pour lesquels il existe des compétences nationales."

9. Burkina Faso, 1991, p. 54.

10. Chad, "Etude Diagnostique de la Coopération Technique," février 1990, p. 38.

11. UNDP background document for CAR Round Table, Geneva, April 1991.

12. M. Maiga, "Etude de Cas sur l'Assistance Technique dans le Secteur Santé," Gouvernement du Mali/PNUD, octobre 1990.

13. République de Burundi, "Rapport du 1ère Séminaire National sur l'Assistance Technique," septembre 1988.

14. See NaTCAP Burkina Faso, "Présentation Succincte à l'attention de M. le Ministre du M.P.C.," novembre 1990. See also, NaTCAP diagnostic study, Burkina Faso, "Contraintes à la Relève," septembre 1990. The NatCAP papers provide further details. In agriculture, there were 73 technical assistance personnel in the country while 58 Burkinabe with university degrees in agriculture were unemployed and another 200 were being trained in the country and abroad. In livestock, the country had 10 technical assistance personnel, 36 unemployed Burkinabe, and 45 in training. In hydrology, there were 33 technical assistance personnel, while 13 Burkinabe were unemployed and 24 were in training.

15. Chad, 1990, p. 22. By contrast, 81 percent of the respondents reported that more training was needed in their projects and 75 percent expressed desire for more equipment.

16. These terms are used interchangeably here and throughout, although this is an oversimplification. Chapter Two goes into the distinction between the two terms.

17. See, for example, République de Burundi, "Document d'Orientation sur la Coopération Technique," 1990; Burkina Faso, "Coopération Technique, Document d'Orientation," 1991.

18. Ghana, "Technical Cooperation Policy Framework Paper," 1989, p. 4.

19. Uganda, "Technical Cooperation Policy Framework Paper," 1992.

20. A Lesotho study notes: "Lack of counterparts, or the removal of trained counterparts through promotion, transfer . . . or . . . 'brain drain' to the S. African homelands, the parastatal sector or private business is probably the most common project implementation problem as perceived by the donor agencies." Government of Lesotho, "Technical Cooperation Paper: Round Table Conference for Lesotho," October 1988.

21. See, for example, Rita Cruise O'Brien, "Colonization to Co-operation? French Technical Assistance in Senegal," *Journal of Development Studies*, October 1971.

22. See, for example, Chapter Six of Robert Cassen and Associates, *Does AID Work?* Oxford: Clarendon Press, 1986.

23. Examples are the 1986 Nordic Conference on Technical Cooperation; the 1986 Development Assistance Committee of OECD meeting on Reassessing the Role of Technical Assistance in Strengthening Public Management Capacities in Low Income Countries; World Bank Seminars on the Management of Technical Assistance in Nairobi in 1987 and 1992 and in Berlin in 1988; the Cluster Meetings of African Ministers of Planning organized by UNDP in November and December 1988; the U.N. Interagency Task Force meetings on technical cooperation in 1989; the 1989 German Development Foundation Round Table on the Future of Technical Assistance; many Development Assistance Committee (DAC) meetings between 1989 and 1992 for discussion of the report on new principles for technical cooperation; the October 1991 conference on New Directions in Technical Cooperation in Africa sponsored by the European Center for Development Policy Management at Maastricht; and the 1991 Conference on Technical Cooperation and the NaTCAP Experience organized by German Development Foundation and UNDP.

24. "Technical Cooperation in African Development: An Assessment of its Effectiveness in Support of the United Nations Programme of Action for African Economic Recovery and Development, 1986-1990," UNPAAERD, UNDP, March 1989.

25. Beatrice Buyck, "The Bank's Use of Technical Assistance for Institutional Development," Working Paper Series no. 578, The World Bank, January 1991, p. v. Annex 1 of this report contains a candid critique of the problems of World Bank-sponsored technical assistance, drawing heavily on African examples.

26. Cited in European Center for Development Policy Management, "New Avenues for Technical Cooperation in Africa," Maastricht, 1992.

27. Richard Jolly, "A Future for UN Aid and Technical Assistance?" *Development 89*, Journal of SID, no. 4, p. 21.

28. According to the DAC 1991 report already cited: "Technical assistance has been for more than thirty years one of the pillars of official development assistance. It has had extremely positive effects, for example, the training of a very great number of nationals from recipient countries and the strengthening of many institutions throughout the developing world." The report goes on, however, to note technical cooperation's weaknesses and failures: "At the same time, however, the tool so long used as the solution to many problems has tended to become a problem in itself. Some of the failures are due to the extremely difficult conditions in which TC [technical cooperation] takes place. TC can only be as effective as the policies and receiving structures of the recipient, and it is the very function of Technical Cooperation to strengthen them. Beyond these intrinsic difficulties, there are still too many cases of inadequate planning and management, lack of careful and realistic definitions of objectives, over-emphasis on project and implementation requirements, over-reliance on expatriate experts and under-use of local expertise, and duplication and competition among donors."

29. Beatrice Buyck, "Technical Assistance as A Delivery Mechanism for Institutional Development: A Review of Issues and Lessons of Bank Experience," Country Economics Department, Public Sector Management and Private Sector Development Division, World Bank, December 1989, p. 18.

30. According to a new school of evaluation theorists, reality can never be "objective." The most important data in evaluations are the perceptions of key stakeholders because behavior is based not on reality itself but on perceptions of reality. The purpose of an evaluation, therefore, should not be to arrive at some ultimate rendition of reality as perceived by outside evaluators, but to reconcile the different views of reality that may be held by various stakeholders. Egon Guba and Yvonna S. Lincoln, *Fourth Generation Evaluation*, Newbury Park, California: Sage Publications, 1989.

31. Robert J. Muscat, "Evaluating Technical Cooperation: A Review of the Literature," *Development Policy Review*, London: Sage Publications, 1986.

32. All citations in this section are from Cassen and Associates, 1986, Chapter Six.

33. Kim Forss, J. Carlsen, E. Froyland, T. Sitari, and K. Vilby, "Evaluation of the Effectiveness of Technical Assistance Personnel Financed by the Nordic Countries," 1990.

34. Ibid., p. 23.

35. Ibid., p. 26.

36. Ibid., p. 56.

37. The Forss report found that of the 324 technical assistance personnel interviewed, 65 percent were "implementers," 17 percent were "controllers," 11 percent were "trainers," and only 7 percent were "institution builders." Ibid. p. iii.

38. Ibid., p. ii.

39. Ibid., p. 80.

40. During the 1980s, World Bank technical assistance credits to Africa have averaged about $360 million, constituting about 15 percent of all lending to Africa or about 25 percent when adjustment lending is excluded. World Bank, "Bank-Financed Technical Assistance Activities in Sub-Saharan Africa," May 2, 1989, p. 11.

41. About 70 percent of World Bank technical assistance is now of the soft variety. World Bank, *Finance and Development*, December 1990, p. 27.

42. World Bank, Operations Evaluation Department, "Evaluation Results for 1988: Issues in World Bank Lending Over Two Decades," 1990, p. 6.

43. Included in the World Bank's definition of technical assistance is project-related training including classroom, overseas fellowships, seminars, and on-the-job training; the technical assistance components of investment projects as well as free-standing technical assistance; and elements of institutional support including the payment of operations costs and local salaries and the provision of minor equipment.

44. Among the most comprehensive reports are "Report on Technical Assistance in Sub-Saharan Africa" (August 1982); "Report on the West Africa Region Task Force on Technical Assistance and Training Effectiveness" (November 1986); "Improving Design and Delivery of Technical Assistance in Africa" (March 1987); "World Bank Technical Assistance Activities and Issues FY 1982-1986" (September 1987); "Technical Assistance as a Delivery Mechanism for Institutional Development: A Review of Issues and Lessons of Bank Experience" (December 1989); and "Free-Standing Technical Assistance for Institutional Development in Sub-Saharan Africa" (April 1990).

45. World Bank, Operations Evaluation Department, "Institutional Development in Africa: A Review of World Bank Project Experience," vol. I, Annex 4, World Bank, 1984.

46. J. Auerhan et al., "Institutional Development in Education and Training in Sub-Saharan African Countries," World Bank Discussion Paper, AFTED, 1985.

47. Page Eaves, "OED Analysis of Institutional Development," paper presented to Conference on Institutional Development and the World Bank, December 1989.

48. World Bank, Operations Evaluation Department, "Evaluation Results for 1990," Washington, D.C., 1991.

49. World Bank, "Managing Technical Assistance in the 1990's: Report of the Technical Assistance Review Task Force," 1991, Annex 6, p. 6.

50. C. Gray, L. Khadiagala, and R. Moore, "Institutional Development Work in the Bank: A Review of 84 Projects," World Bank Staff Working Paper no. 437, 1990. Only 44 percent of these projects are characterized as having "good treatment" of institutional development issues.

51. World Bank, Operations Evaluation Department, "Free-Standing Technical Assistance . . . ," 1990, p. iii. It is illustrative of the methodological difficulties presented by evaluations of the effectiveness of TA/ID that at least one project that was found to be successful in the World Bank study was found to be far less successful by another study. See Elliot Berg, "The Reform of Public Investment Programming in Senegal: An Evaluation," Development Alternatives, Inc., Bethesda, Md., October 1991.

52. World Bank, "Managing Technical Assistance in the 1990's . . . ," 1991, Annex 4.

53. The evaluation method used was to review project completion reports and Evaluation Department audits and, in some cases, to discuss the evaluations with the evaluator. Projects were then rated on a scale of one to three, one for negligible results, two for partial, and three for substantial. On this basis, an effectiveness index was calculated. The overall rating worldwide was 1.69. The rating for African projects was 1.5.

54. The equipment component of technical cooperation is also frequently faulted, on two counts. The first is that the vehicles, computers, copiers, and so forth come from so many different donors and are therefore of so many different types that maintenance is complicated and spare parts shortages magnified. The second criticism is that little or no provision is made for maintenance of the equipment once the technical cooperation personnel have departed or the training has been completed.

55. "Management" in this context refers to management of the entire cycle of technical cooperation activities including policy formulation, planning and programming, project design, implementation, and evaluation.

56. See Kjell J. Havnevik, "Unequal Partners: The Role of Donors and Recipients in the Identification, Design and Implementation of Technical Cooperation Projects," UNDP, May 1990.

2

DEFINITIONS, HISTORY, STATISTICS

Technical cooperation—or its rough equivalent, technical assistance—seems a simple enough concept. Stop 10 people on a street anywhere and ask them what they understand it to mean, and you would probably get the essential ingredients of a reasonable definition: that it is a form of foreign aid, that it involves sending experts from richer to poorer countries to supply missing skills and train local people, and that it has something to do with introducing better ways of doing things.

If 1 of the 10 respondents happened to be especially learned, she might say that its essence is training and the building of institutional competence. This comes close to an acceptable brief definition, but it still leaves much unsaid. There is more to the concepts of technical cooperation and technical assistance than meets the eye, and that these two terms are given different meanings by different members of the development community is a source of considerable confusion. It is important, therefore, to sort out these definitional matters, and to make clear what we and others are talking about when these terms are under discussion. That is the substance of the first part of this chapter.

The second part of the chapter looks backward. For 30 years, or since most of its states became independent, Sub-Saharan Africa has been the recipient of more technical cooperation than any other region. Most of the reasons for this assistance derive from the colonial situation and its relatively recent demise in Sub-Saharan Africa.

The third part of the chapter is a quantitative analysis of the present size and nature of technical cooperation flows into Sub-Saharan Africa. It summarizes available data on trends in technical cooperation—its financial magnitude, origin, country distribution, and sectoral allocation.

The chapter is thus mainly prelude. It tries to make clear what the authors mean by technical cooperation and explain its extensive presence in Sub-Saharan Africa. It also sets out the basic statistics showing how much technical cooperation there is, what sectors it works in, where it comes from, and where it goes. All of this is useful background for understanding the spreading discontent with this form of aid that has marked the late 1980s and the ensuing calls for change.

CONCEPTS AND DEFINITIONS

Although the essentials of technical assistance or technical cooperation might be easily grasped, numerous complexities surround the concepts, and confusions have grown up around their use. Partly this is due simply to loose use of language. But historical and bureaucratic factors also enter, because each development agency has endowed technical cooperation and related concepts with special meanings.

Definitional complications have developed around distinctions between several key concepts:
- Technical assistance and technical cooperation;
- Technical cooperation and capital assistance;
- Investment (or project-related) technical assistance and free-standing technical assistance; and
- Technical assistance for operational support and for institutional development, or capacity building. (Different usages of

institutional development and capacity building give rise to additional uncertainties.)

The Distinction Between Technical Assistance and Technical Cooperation

In everyday language technical assistance most commonly refers to the provision of personnel by a donor to a recipient. Technical cooperation sometimes means the same thing but in some cases refers to a broader set of development activities including training, information exchanges, and the supply of equipment and materials. (Box 2:1 shows various definitions of technical assistance and technical cooperation.)

This distinction has faded in recent years; the terms are now in most cases used interchangeably, although some donors continue to define them as distinct concepts. Technical assistance was the term more commonly used until the 1970s when technical cooperation came into use. The U.N. system adopted this term, although technical assistance is also sometimes used by U.N. staff in routine discussions. The World Bank, A.I.D., and the Canadian International Development Agency (CIDA) use the term technical assistance rather than technical cooperation, although the latter is coming into more general use. The French government uses the term *assistance technique* to mean provision of long-term expatriate personnel.[1] One semantic advantage of the term technical cooperation is that it conveys the notion of a partnership between donors and recipients, whereas technical assistance connotes a giver-taker relationship.

In many donor development agencies, specialized units have been created for the administration of technical cooperation and technical assistance and programmes—for example, GTZ in Germany and the Japan International Cooperative Agency. Whatever the nature of activities or inputs managed by these units, their projects or programmes tend to be classified automatically as technical cooperation or technical assistance.

BOX 2.1

DEFINITIONS OF TECHNICAL COOPERATION AND
TECHNICAL ASSISTANCE

- The following definitions are used by the Organization for Economic Co-operation and Development for statistical collection purposes, and included in the DAC/Working Party's guidelines.

    Technical cooperation: Activities whose primary purpose is to augment the level of knowledge, skills, technical know-how, or productive aptitudes of the population of developing countries, i.e., increasing their stock of human intellectual capital or their capacity for more effective use of their existing factor endowment.

    Technical assistance: Financing of services with the primary purpose of contributing to the design and/or implementation of a project or programme aiming to increase the physical capital stock of the recipient country.

    Source: *Geographical Distribution of Financial Flows to Developing Countries*, OECD, Paris, 1991.

    The following definition is provided by the DAC "Principles for New Orientations in Technical Cooperation," Paris, 1991.

    Technical cooperation (TC) encompasses the whole range of assistance activities designed to improve the level of skills, knowledge, technical know-how and productive aptitudes of the population in a developing country. A particularly important objective of Technical Cooperation is institutional development, i.e. to contribute to the strengthening and improved functioning of the many institutions essential for sustainable development through the effective management and operation of an economy and of society more generally.

- World Bank reports on technical assistance, and related documents and studies, use various definitions. The following is the most recent definition, set down in a review of Bank technical assistance programmes.

    Technical assistance is defined as the transfer or adaptation of ideas, knowledge, practices, technologies, or skills to foster economic development. The purposes of Bank technical assistance are classified as follows: (a) Policy development, (b) Institutional development, (c) Capacity building, and (d) Project or programme support.

    Source: "Managing Technical Assistance in the 1990s," Report of the Technical Assistance Review Task Force, World Bank, October 1991.

BOX 2.1 — Continued

- UNDP uses the following definition in its guidelines for statistical collection in its Development Cooperation Information System.

    Technical cooperation comprises the provision on concessionary terms of resources aimed at transfer of skills and know-how and at capacity building within national institutions to undertake development activities. It includes resources in form of personnel (international, national, long, short term). Technical cooperation may be broadly divided into two categories: (a) investment related, or technical cooperation inputs necessary to assist in the implementation of capital investment projects, and (b) general institutional support, or "free-standing" technical co-operation which is provided regardless of the needs of specific investment projects.

Source: UNDP. *NaTCAP Methodology*. Regional Bureau for Africa, May 1989.

The following is the definition that sets out the purpose of technical cooperation to guide UNDP in its operational programmes. This definition was developed in 1975 as part of a major review of the purpose of technical cooperation in the UN system, and the need to set policies in line with the requirements of developing countries:

   The basic purpose of technical cooperation should be the promotion of self-reliance in developing countries, by building up, inter alia, their productive capability and their indigenous resources—by increasing the availability of the managerial, technical, administrative and research capabilities required in the development process.

Source: UNDP Governing Council Decision 75/34, "New Dimensions in Technical Cooperation," June 1975

Another UNDP statement lists the following types of activities as constituting technical cooperation:

— Human resource development through transfer of skills and know-how, training in institutions and on the job, and activities in social fields such as health, nutrition, education;

— Preparation of development plans and strategies and of feasibility studies and the acquisition of basic information and data;

— Transfer, adaptation, development, and diffusion of technology;

— Development of services such as administrative services; and

— Development and execution of pioneering programmes.

BOX 2.1 — Continued

Furthermore, it is noted that "the modern concept of technical cooperation should be seen in terms of the above-mentioned and similar achievement categories and not in terms of specific input or budget categories. Programmes of technical cooperation should be able to devise and utilize new methods, processes and institutional responses to meet the more sophisticated needs of the 1970s and of the decades to come. They should be flexible enough to respond to a variety of needs."

Source: UNDP, "The Future Role of UNDP in World Development in the Context of the Preparations for the Seventh Special Session of the General Assembly Governing Council document DP/114," March 24, 1975, New York.

OTHER DEFINITIONS

- Broadly speaking, <u>any activity aimed at enhancing human and institutional capabilities</u> through the transfer, adaptation and utilization of knowledge, skills and technology, can be considered as TC. It refers to ODA financed expatriate personnel (experts, volunteers, consultants), students and trainees and a wide range of activities and services (feasibility studies, engineering and construction services for capital projects, institution building efforts, transfer of managerial skills, research related to development, equipment and supplies, etc.).

  Source: European Center for Development Policy Management, *New Avenues for Technical Cooperation in Africa*, Maastricht, 1991.

- A <u>range of activities</u> that enhance and/or complement <u>human and institutional capabilities</u> through the development, transfer, adoption and the use of skills and technology from sources external to the government/recipient agency.

  Source: Beatrice Buyck, *The World Bank's Use of Technical Assistance for Institutional Development. A Review of Issues and Lessons of Experience*, 1989.

- Activity to promote socioeconomic development by enhancing <u>human and institutional capacity</u> and by transfer of knowledge and technology. In other words, technical cooperation aims to build self-reliance in developing countries.

  Source: Pierre-Claver Damiba, address to the Seminar on National Technical Cooperation Assessment and Programmes (NaTCAP), Lilongwe, Malawi, September 7-9, 1989.

The Development Assistance Committee of the Organization for Economic Co-operation and Development (OECD) defines technical cooperation and technical assistance as two distinct categories of aid; technical cooperation is limited to free-standing activities aimed at institutional development or capacity building, while technical assistance is defined as support to implement investment projects.[2] The World Bank makes the same kind of distinction between investment-related and free-standing assistance situations, but uses the term technical assistance to cover both.

A major difficulty with the OECD definitions of technical cooperation and technical assistance is that the two categories are not—or at least should not be—mutually exclusive. Many investment-related technical assistance activities have as their prime objective implementation of a capital project. But they also contribute to creating human capital. Expatriate teams may come to a country to carry out feasibility studies, provide management services to run enterprises, or set up a factory on a turnkey contract. But, even in these cases, there is usually some contact with local personnel and, hence, some training benefit from these contacts. Moreover, training will eventually be needed to allow efficient and sustainable use of these capital investments.

In this book, technical cooperation and technical assistance are used synonymously and are defined as indicated earlier: activities that augment and improve human or institutional resources. The term technical cooperation is given preference. Instead of technical assistance defined as transfer of expertise, we will refer to technical assistance personnel; this term will include all personnel financed by donors—expatriate "experts," "consultants," and "volunteers," as well as "national experts and consultants." The conceptual problems raised by inclusion of the latter group have already been mentioned and are discussed in greater detail in Chapter Six.

The Distinction Between Technical Cooperation and Capital Assistance

Development assistance takes three main forms: capital assistance; technical cooperation; and budget or balance of payments support, nowadays usually conditioned on policy reform. What is distinctive about technical cooperation, especially with regard to capital assistance?

At first blush the answer seems easy. Technical cooperation provides expertise and training, and capital assistance provides physical infrastructure and equipment. This perception—that these forms of aid differ mainly by the nature of the inputs they provide—is widely held. It seems to be in line with what is observable in day-to-day operational reality because technical cooperation normally takes the following forms: (1) the provision of expatriate personnel to carry out any activity related to development; (2) assistance for implementation of capital projects; (3) organizational reinforcement through provision of management and other operational or advisory personnel; (4) provision of advisers in policy analysis and formulation; (5) training; and (6) research and studies, including feasibility studies for capital projects.

However, to distinguish between technical cooperation and capital assistance only on the basis of inputs is analytically dubious. The distinction ignores the objectives of technical cooperation, clearly an incorrect approach. After all, the objectives of any activity should determine the inputs needed to achieve it. The objectives of technical cooperation are to facilitate the acquisition of skills and know-how and to develop national capacity for managing development. Achievement of these objectives may require inputs other than expatriate personnel, training, and the supplies and equipment necessary to allow these personnel to do their work and to implement training programs. It may be necessary, for example, to acquire and use specialized equipment or to undertake policy research in pursuit of technical cooperation objectives. Capital investments to improve the work environment may also be necessary; who, after all, can learn new techniques or procedures expeditiously when five people are crowded in a small office, air conditioning systems are down, windows are broken, and equipment of all kinds is scarce and frequently inoperative.

Help for general operating budgets may also be called for: how can government agencies become stronger when gasoline, paper, typewriter ribbons, and light bulbs lack, when budget support for maintenance and spare parts evaporates, when electricity bills go unpaid, when vehicles are few and vulnerable, budgets for fuel are tiny, and when even money for stamps and paper may be hard to find.

Thus, the distinction between capital assistance and technical cooperation on the basis of types of inputs provided is blurred. Improved physical infrastructure (buildings, equipment) and added operating costs are often helpful in transferring skills and important for reinforcing the capacity of national institutions. Therefore, both technical cooperation and capital assistance can involve the use of similar inputs, although in different proportions.[3]

Other areas of definitional ambiguity exist. Because of fiscal stress, many African governments accept expatriate technical assistance personnel as substitutes for local expertise, which is available but which governments are unable to hire. In this case the distinction between technical cooperation and general budget support becomes hazy. By financing the hiring of expatriates for jobs that locals could do if domestic budget resources were available, the assistance takes on the character of sub rosa budget support.

It is possible to set out a cleaner set of analytic distinctions that takes into account both inputs and objectives or outputs. Technical cooperation is a distinct type of development assistance, distinguishable from aid in the form of capital investment (or aid in other forms, such as budget or balance of payments support). Each of these forms of aid is aimed at increasing output and rates of growth, although technical cooperation has an additional objective: increasing self-reliance. Although there is some overlap, each reaches its objective by a different path: technical cooperation by raising general factor productivity and building local capacity, capital assistance by increasing the supply of capital, and policy-conditioned budget or balance of payments lending by changing the policy environment.

The proximate objective of technical cooperation is to raise productivity of the existing stock of factors of production in the recipient economy. The ultimate objective is increased self-reliance—by transferring know-how and skill and by increasing local capacity to manage national resources.

Technical cooperation raises productivity in two main ways. First, technical cooperation augments the stock of human capital—temporarily, through importing skills unavailable locally (filling skill gaps) and, more permanently, by transferring skills and technical know-how to local people. Second, technical cooperation increases the effectiveness of local institutions, by introducing better administrative and other procedures. Other inputs (materials and supplies) that are often part of technical cooperation reduce key constraints on the effectiveness of public sector employees and hence are also productivity enhancing.

Capital assistance involves an increase in the supply of physical capital. It finances infrastructure—roads, dams, and irrigation canals—as well as factories and durable equipment. An increased stock of physical capital should lead to increased productivity of labor and land, higher output, and faster growth.

These simple distinctions capture the main elements of the technical cooperation concept and indicate its main points of difference with capital assistance. Capital assistance is understood to mean the financing of brick and mortar or capital equipment, with the primary aim of increasing output and growth. Technical cooperation finances mainly an intangible input—knowledge—that also raises output but aims, in addition, at increased competence of national human resources and institutions. The material instruments for this knowledge transfer are technical assistance experts, training programs, and other inputs needed to do things better and apply new technology.

This kind of analysis serves to clarify understanding about what is and is not distinctive about technical cooperation as compared with other forms of development assistance. But in the real world such refinements count for little. Bureaucratic tradition and history often determine how aid activities are classified and hence how aid expenditures are categorized. In practice, there is much that is arbitrary about how aid activities are classified. This is true, for example, of feasibility studies and other preinvestment actions. They can be classified as investment or as technical cooperation. They are mostly included as investments in state development budgets and public investment programs. If the definition of investment projects is narrowed to include only capital investment, these studies should be classified as technical cooperation. Similarly, when run-down, barely habitable office buildings are replaced, the expenditure should be classified as capital investment. But such activities are part of some technical cooperation projects and classified as technical cooperation expenditures.

For other kinds of projects or activities, the mix of inputs determines its classification. This is the case with the technical cooperation projects of UNDP; its operating rules prevent expenditure of more than 50 percent of total project costs on equipment or operating costs.[4]

In any case, the reality is that projects or programmes called technical cooperation vary with respect to input composition. For example, French government technical cooperation refers, in practice, strictly to technical assistance personnel with very limited accompanying equipment such as vehicles and operating materials. UNDP technical cooperation projects give much bigger scope for other inputs; they explicitly recognize the possibility of using national expertise and equipment for capacity building, and a few UNDP-financed projects, particularly in Asia and Europe, consist mainly of high-technology equipment. Some World Bank free-standing technical assistance projects also have very large equipment and building components. In the case of many bilateral donors—Canada, the United States, the Netherlands, and Sweden, among others—there is no special category of technical cooperation projects as distinct from capital projects.[5]

Because so many writers and practitioners use technical cooperation-related terminology in rough-and-ready fashion, it is not surprising that loose definitions are common or that individuals and aid organizations use these terms to mean different things in different contexts. This has led to difficulties, particularly for statistical-collection purposes; there is much that is unclear about what is counted as technical cooperation and about comparability of data provided by different donors and recipient countries.

Should, for example, financing of local expertise be counted as technical cooperation or as financial assistance? Should equipment, training, and operating costs be included as technical cooperation? Should long-term personnel assigned to help manage a capital assistance project be counted as technical cooperation personnel? How about volunteers from nongovernmental organizations such as church groups? There are no cut-and-dried answers to these and many other questions, and because donor and recipient country responses to them often vary, there is some risk of misunderstanding in policy debates, and the statistical database on technical cooperation is often vague and uncertain.

In this book we follow the general definitions outlined above. The term capital assistance will refer to programs and projects that aim principally at creating physical assets. The terms technical cooperation or technical assistance will refer to programs and projects that aim mainly at increasing or improving human or institutional resources. Both capital assistance and technical cooperation can be viewed as investments in the sense that both should achieve an increase in the stream of future output, although by somewhat different paths. And technical cooperation has an additional objective—that of increasing self-reliance through strengthened local capacity to manage resources.

Distinction Between Investment-Related and Free-Standing Technical Assistance

Many donors make this distinction. In World Bank parlance, investment-related (or project-related) technical assistance means technical assistance provided within the framework of an investment project (really a capital project), while free-standing technical assistance is provided for institutional development or, more broadly, for any purpose not directly related to an investment (capital) project. (Free-standing assistance is sometimes called nonproject assistance.) Bank documents also call investment-related technical assistance "hard," as compared with the "soft" technical assistance that goes to institution building. Hard technical assistance involves engineering and other scientific know-how for such purposes as feasibility studies, drawing up of bidding documents, and supervision of project construction. Soft technical assistance involves training and provision of technical assistance personnel for such purposes as strengthening management information systems, helping in policy formulation, and conducting research.

Another distinction between investment-related and free-standing technical assistance is that the former usually comes with some equipment included in the project, while the latter does not. In Francophone countries, free-standing assistance is unofficially referred to as *assistance technique tout nu*.

As noted earlier, DAC of OECD makes a similar distinction and distinguishes technical assistance and technical cooperation along the same lines. Thus technical assistance is defined, in DAC documents, as inputs of expertise that are part of a capital project, while technical cooperation is the name given to the technical assistance personnel and other inputs that are used in free-standing activities aimed at training and institutional development.

The DAC definition of technical cooperation is particularly important because it forms the basis of the comprehensive crossnational database on technical cooperation that is published in DAC's annual report on development cooperation. Because DAC does not include "investment-related technical assistance" in its official definition of technical cooperation, technical assistance and training that are provided as components of investment-related projects are excluded from the OECD database.

Distinction Between Operational Support and Institutional Development Technical Cooperation

Some donors and many students of foreign assistance have made other attempts at categorizing technical cooperation, focusing particularly on distinctions between technical cooperation activities that are aimed essentially at "getting the job done" and those intended to transfer technology, train nationals, and develop sustainable capacity. These two broad categories are most commonly referred to as operational or direct-support technical cooperation, on the one hand, and institutional development technical cooperation, on the other.

Personnel can be distinguished along similar lines. There are "doers" and "advisers/trainers." The "doers" are said to be filling substitution functions; they are referred to as "substitution technical assistance," or sometimes as "performer technical assistance." Substitution personnel historically occupied line posts or were integrated into the national administrative hierarchy. The French system of *coopérants* or the British topping-up arrangements and the provision of doctors and teachers by some non-OECD bilateral cooperation were specially designed to provide this assistance. The justification for this type of assistance is gap-filling, or providing personnel when qualified nationals are not available.

Advisers and trainers are brought in to provide training, in-service and on-the-job, to target groups of local staff. Their basic task is skills transfer.

In principle, advisers and trainers have become the predominant form of technical assistance personnel. Provision of technical assistance personnel to fill line posts is much less common today than it was in the immediate post-independence period. As noted earlier, however, one major problem of technical cooperation today is that, regardless of their formal mandate, many technical assistance personnel appear to continue to perform such substitution functions; they do the job themselves rather than train counterparts.

Some complications have been introduced into these definitional matters by translation problems. For example, much of the French technical cooperation programme has consisted of placing *coopérants* at the disposal of recipient governments to perform substitution-type technical assistance. Personnel are assigned as individuals, not within the context of a project. They are given general mandates. (Since 1991, France's new policy has been to phase out this form of assistance in favor of personnel with more specialized expertise provided to assist in specific activities defined in detailed terms of reference [*lettre de mission*]). Because the term "free-standing" defies translation into

French, it is sometimes mistakenly associated with substitution-type technical assistance.

Project-related technical assistance for the French refers to the assignment of personnel as part of a development project, which can be either investment related or for institutional development. The key distinction is that the project-related technical assistants have specific and time-bound objectives, whereas free-standing technical assistance personnel do not.[6]

In addition to substitution personnel, the literature on technical cooperation refers to other useful categories. Operational support includes "control technical assistance personnel": personnel assigned by donors to play a direct role in technical, administrative, or financial management of projects. Institutional development technical assistance personnel perform functions of adviser, teacher, catalyst, or mobilizer.[7] Box 2.2 shows various classification schemes for technical assistance personnel.

Underlying many of these definitional ambiguities lies a contradiction. Almost all development activities require both capital investment and human and institutional capacity. Yet habits of thought and bureaucratic traditions tend to treat them differently. The conception of technical cooperation as a separable activity rests on an artificial division: it splits off the human and institutional dimension and couples the objectives of capacity building with particular inputs such as expatriate personnel and training.

This was the historical conception: that the output or expected result of a capital assistance project was some type of physical asset (usually infrastructure), while the expected result of a technical cooperation project was enhanced human or institutional resources. Some aid agencies tended to specialize in one or the other kind of development activity. But the fact is that most development projects require both types of inputs and aim at both types of outputs.

A similar point can be made for the impact of development projects over time. For development to be sustainable, management and human and institutional factors are as indispensable as capital investment.

Along the same lines, the distinction between direct support and institution-building technical cooperation is not or should not be

BOX 2.2

SOME CLASSIFICATIONS OR TYPES OF TECHNICAL ASSISTANCE

Following are some overlapping classifications of technical cooperation that are useful for analytic purposes.

A. Classified by Overall objective

    1. Hard
Focuses on accomplishment of a specific, concrete task with a measurable output

    2. Soft
Focuses on institutional and human resource development and training

B. Classified by Relationship to a Project

    1. Project-Related
Technical assistance as a component of an investment project that focuses primarily on physical outputs

    2. Free-Standing
Technical assistance assigned to help strengthen an institution or support policy adjustment without any link to a physical investment project

C. From Point of View of Donor

    1. Skills Transfer
Objective being for the technical assistance to transfer skills to a counterpart or counterparts

    2. Control
Objective being for the technical assistance to manage or control the use of project resources

    3. Catalyst
Objective being for the technical assistance to breathe life into a dormant project or institution

BOX 2.2 — Continued

D1. Technical Assistance Personnel Classified by Type of Task Performed

   1. Performer
      Engaged principally to perform a task or deliver an output

   2. Substitute
      Temporarily taking the place of a national who is away for training

   3. Teacher
      Engaged principally to transfer skills through formal or informal training

   4. Mobilizer
      Engaged principally to create or strengthen an institution through consciousness raising, organization development, training, and coalition building

D2. A recent World Bank effort to classify technical assistance in Mozambique by the type of task performed by the expatriate gave the following categories:*

   (a) **Foreign experts** are high-level policy advisers to the government. They can be either resident advisers or specialists, brought in for either short-term or long-term assignment, to perform tasks for which no local expertise exists;

   (b) **Gap-fillers** are expatriates that are brought in to fill already established positions within the government or parastatal sector. In the Mozambican context, these workers are most often known as *coopérantes*. In general, gap-fillers perform tasks that could be filled by Mozambicans, but the combination of labor scarcity, poor incentives in the civil service, and, in some cases, donor-driven hiring practices has created a situation in which *coopérantes* have become an established group within the Mozambican civil service;

   (c) **Gatekeepers** are expatriates hired by donors with the specific task of managing project implementation; and

   (d) **Institutional twins** are foreign institutions that have been contracted on a long-term basis to engage in institutional development with a homologue in Mozambique.

* This classification comes from the World Bank, "Capacity Building in Mozambique," 1991. It draws on the typology set out in John M. Cohen, "Expatriate Advisors in the Government of Kenya: Why they are there and what can be done about it," Harvard Institute for International Development, Development Discussion Paper No. 376, June 1991. Cohen focuses on long-term advisers to the civil service, and distinguishes five types of adviser, by function: high-level, gap-filler, condition precedent, gatekeeper, and specialist.

watertight. The two categories of technical cooperation are not mutually exclusive. Many direct support activities could and should have an on-the-job training impact on the national personnel involved in these donor-financed projects, and on institutional effectiveness as well.

Institutional Development and Capacity Building

Recent international debate on technical cooperation has focused attention on the importance of institutional development or capacity building—the strengthening of sustainable indigenous capacity to manage economic change and growth. There is a growing consensus that the development of this capacity should be the central, if not the unique, goal of technical cooperation. In its "Principles for New Orientations in Technical Cooperation," DAC states:

> Progress towards sustained, more equitable and self-reliant development depends critically on the strength and quality of a country's institutional capacity. Contributing to this objective must therefore be an essential purpose of development co-operation in general and Technical Co-operation in particular. An aid activity cannot be regarded as successful unless it has contributed to strengthening the local institutions through which and for which it works.[8]

A striking reflection of this preoccupation with capacity building is the creation in 1991 of the African Capacity Building Foundation, a joint donor-African government entity, which has an initial endowment of almost $100 million. The underlying idea that motivates the new organization is set out in a recent report.[9]

> A vital gap is not being adequately filled: capacity in economic policy analysis and development management. Despite the achievements in education and training in Africa during the past 30 years, most countries still do not have a critical mass of top-flight policy analysts and managers who can help pilot their

economies through the storms and turbulence that must be faced daily.

The policy statements on technical cooperation prepared in over a dozen African countries in recent years also emphasize capacity-building aspects of this form of aid. The new orientation, as observable in government statements on technical cooperation policy, emphasizes the primacy of the capacity-building objective. The following are typical statements:
- Ghana's 1990 TCPFP defines technical cooperation as "that aspect of international economic cooperation aimed at promoting the socio-economic development of developing countries through the transfer of technical knowledge and the development of human resources and institutions";
- Malawi's 1990 policy document is entitled "Statement of Policy on Technical cooperation for Human and Institutional Development." Its foreword states: "Government considers its technical cooperation programmes, carried out in close collaboration with the donor community, as a major means for building up the country's institutional and human capabilities to execute national development programmes";
- The policy orientations of the government of Chad, as laid out in a paper dated June 1990, states as one basic principle that technical cooperation has to emphasize training and transfer of know-how; and
- The Burkina Faso policy orientation document of November 1991 states that the ultimate objective of technical cooperation is to reinforce national capacity, and defines several specific measures to ensure that all technical cooperation activities explicitly address the objective of transfer of skills and training of national counterparts.

Reasons for this renewed emphasis on institutional development and capacity building are not hard to find. Perhaps the clearest lesson of the past 30 years of development effort is that weak performance of indigenous institutions is one of the biggest bottlenecks to faster, better-quality economic growth—maybe the biggest bottleneck. The

World Bank's long-term perspective study put it this way: "Africa's lack of technical skills and strong public and private institutions accounts more than anything else for its current predicament."[10] Everybody knows now that the "hardware" dimension of development—the physical infrastructure, for example—is a lot easier to put in place than the "software" to keep it operable, which depends on local skills and institutions.

Another factor is that expansion of African educational systems has altered the nature of skill development problems; it is now less a matter of filling posts with nationals who have suitable formal qualifications than of improving the qualitative performance of staff and making such performance sustainable.

This focus on capacity building and institutional development gives urgency to the question of what exactly these terms mean and how they differ, if at all. This turns out to be no easy task. The general definition of capacity building has not yet attracted much analytic attention, perhaps because the term has come into common use so recently. And no general consensus has been achieved about the exact meaning of institutional development, despite a complex literature on the subject that dates back at least 40 years. There appears, nevertheless, to be a modicum of agreement on certain key ideas.

Institutions and Organizations

In everyday language, the terms "institution" and "organization" are often used synonymously. But most social scientists distinguish the two. An institution is variously defined as "a pattern of behavior that is valued within a culture," or a "sanctioned norm of conduct or rule of the game that guides and constrains individual and group behavior" (such as marriage, private property, or kinship obligations). In a few cases, where they are particularly deeply rooted, organizations are identified as institutions: cases in point are the Church of Rome, the Sorbonne, and the British Treasury.[11]

In most cases, organizations differ from institutions. An organization can be described, generically, as the rational coordination of activities by a group of individuals with the aim of achieving some common purpose.[12]

Two points emerge. First, organizations form part of the fabric of institutions but are not institutions themselves. Organizations can, moreover, be changed or even eliminated without dramatically affecting an institution. A private grocery store is an organization that contributes to the institution known as private enterprise, but if the store goes bankrupt, private enterprise will endure. A particular ministry forms part of the organization of the institution known as government, but if that ministry is abolished, government will carry on. Consequently, organizations might come and go rather easily, institutions less easily. But when institutional change does occur, it has a much deeper and longer-lasting effect on society.

Second, institutional development means more than just structural or functional changes of an organization. It involves fundamental social change, the transformation of patterns of behavior that are specific to the society. Institutional development is a much more profound process than organizational development.

This is not always clear from what is written about institutional development, or indeed from the actions that aid agencies pursue in its name. Thus some studies define institutional development largely as organizational development. One of them sees it as "the process of improving the ability of (organizations) to make effective use of the human and financial resources available."[13] But institutional development encompasses more than is suggested in this definition.

Elements of Institutional Development

At least seven different components or dimensions can be discerned:[14]
- Changing the incentive structure for individuals and for organizations to induce personal and bureaucratic behavior that is in greater harmony with development needs;
- Enhancing skills by training and education;
- Strengthening organizational performance;

- Reforming procedures or systems of coordination between organizations;
- Increasing financial capabilities (the command over resources) by more effective mobilization such as better pricing systems, more user fees, and stronger overall revenue generation and use—better planning, budgeting, and expenditure control;
- Nurturing societal supports—for example, by encouraging formation of user groups, political reforms that bring greater transparency and accountability, and greater participation of wage earners in determination of their conditions of employment; and
- Cultivating new norms and values—for example, applause for successful new capitalists in post-Socialist societies, condemnation of corruption, special rewards for managers using participatory styles and delegation of authority, and generally granting honors and awards for socially desirable behavior.

By generous interpretation, many of these could be considered to fall within the scope of the organizational development-focused definition of institutional development given above. But the building of societal supports and the cultivation of new values clearly are outside the scope of such a concept. This would also seem to be true for non-job-specific training and improved procedures such as investment programming.

Technical Cooperation for Capacity Building

What does this analysis of the meaning of institutional development mean for the capacity-building concept and for the orientation of technical cooperation? First of all, the analysis helps give analytic content to the notion of capacity building, which is seen to be something more than organizational development, but less than institutional development defined by its multiple dimensions as above. Capacity building is characterized by three main activities: skill upgrading, both general and job-specific; procedural improvements; and organizational strengthening.

The skill enhancement component includes general education, on-the-job training, and professional deepening in crosscutting skills such as accounting, policy analysis, and information technology. Organizational strengthening covers what some have defined as institutional development—reinforcing the capacity of an organization to use available money and staff more effectively.[15] The procedural improvement dimension refers to functional changes or system reforms, such as introduction of new budgeting arrangements or replacement of ex ante controls over public enterprises by greater autonomy and ex post supervision.

Capacity building by this definition is therefore broader than organizational development in that it includes all types of skill enhancement and also procedural reforms that extend beyond the boundaries of a single organization. Capacity building is less comprehensive than institution building as broadly defined, because capacity building excludes the specific goals of changing social norms and creating societal supports for capacity-raising changes.[16]

Second, this analysis helps define the role of technical cooperation. The primary role of technical cooperation is to build capacity in any of the three key dimensions outlined above: enhancing skills, improving administrative procedures and regulations that condition organizational relationships, and increasing organizational competence. In the rest of this book we will use the term technical cooperation for capacity building to mean technical cooperation directed toward any of these three activities. We will use the term institutional development more sparingly. We recognize that the two concepts are closely related and often used interchangeably in the literature. But institutional development is a much deeper, more complex objective, with dimensions that are beyond the reach of traditional technical cooperation. It is also well to recall throughout that, although technical cooperation is a significant factor in both institutional development and capacity building, it is by no means the only factor, or indeed the most fundamental.

SOME HISTORICAL BACKGROUND

Official technical assistance has its roots in colonial rule and the subsequent decolonization process. It is a common understanding that at the time of independence technical assistance personnel replaced colonial administrators and technicians in the administration. The common assumption was that as the stock of educated labor expanded, the need for technical assistance would decline.

The colonial situation severely inhibited development of technical skills and the emergence of competence in managing large-scale organizations:

- Colonial authorities felt no urgency to push advanced education and were slow even in providing secondary education. In 1958, only 10,000 Africans attended universities, 65 percent of whom were from Ghana or Nigeria.[17] In 1960, there was not a single university in French territories, only two embryonic institutions in Zaire, and a small number of university colleges in some British colonies (Sierra Leone, Ghana, Nigeria, Southern Rhodesia, Uganda, and Kenya).[18]
- Racial barriers, institutionalized and informal, prevented Africans from moving into skilled jobs. Economic incentives driving employers to substitute cheap black for dear white labor or to train blacks on-the-job were frustrated by the social conventions of colonialism: it was taboo for blacks to supervise whites. Tutorial or mentor relationships that might benefit Africans rarely emerged. Furthermore, in the colonial situation, until well into the 20th century, it was easier to import skilled workers to fill shortages than to train Africans.[19] And in colonial public sectors, little sustained effort was made, until the twilight of colonial rule, to train African successors systematically. This was so despite the fact that the financial burden of paying the high costs of colonial administrators absorbed a significant portion of national resources. In Senegal in 1956, for instance, French administrators' wages were 3.2 percent of gross national product (GNP) and one-third of the total civil service wage bill.[20]

In the 1950s, the pace of educational expansion accelerated. The number of Africans attending school, mainly at the primary level, doubled between 1950 and 1960—to about 20 million. But African countries benefitted little from the new technical cooperation programs introduced by bilateral and multilateral donors after World War II.

- The U.K. government disbursed in its colonial territories an average of some 11 million current pounds between 1946 and 1956. Africa received about half this amount, of which only half went to technical cooperation-related activities—education, civil servant training, and research.[21]
- In 1951, a Bureau of Technical Assistance was created within the French Ministry of Foreign Affairs, mainly to supply experts and scholarships for foreign students. The activities of this bureau were limited. In 1953, it provided scholarships for overseas study to 154 students and sent 55 experts overseas for missions ranging in duration from one month to a year.[22]
- In the United States, the newly formed Economic Cooperation Administration offered support only to the few independent states of Africa in the 1950s. Therefore, Africa's share of the available resources was small.
- Africa also failed to benefit significantly from the creation of multilateral technical cooperation sources. The U.N.'s Expanded Programme for Technical Assistance received $19 million in funds pledged for 1952. Of these funds, Africa received only 7 percent. By 1956, at the dawn of the independence period, Africa's share of the approximately $34 million disbursed was only 9 percent.[23]

African states thus came to independence some 30 years ago with very few high-level African technicians, professionals, or managers—people with the training and experience to run modern organizations. Table 2.1 illustrates the high degree of dependence on expatriate manpower that existed in the decade after independence.

Colonial administrations made major efforts to catch up during the last years of colonial rule, but enormous educational tasks remained at independence. In 1960, only 29 percent of African 6-11 year olds, 17 percent of 12-17 year olds, and 1.4 percent of 18-23 year olds were

TABLE 2.1

EXPATRIATE EMPLOYMENT AS A PERCENTAGE OF TOTAL EMPLOYMENT OF TRAINED MANPOWER, c 1967

Country	Year of Survey	Educational Level[a]				Total
		A	B	C	D	
Botswana	1967	94	81	19	42	-
Ivory Coast	1962	79	61	-	-	-
Kenya[b]	1964	77	25	54	1	48
	1969	58	48	36	-	41
Malawi	1966	64	10	14	-	18
Nigeria	1964	39	5	-	-	13
Somalia	1970	7	2	20	2	2
Sudan	1967-1968	12	6	2	0	3
Swaziland	1970	80	74	57	23	35
Tanzania	1965	82	23	31	9	31
	1969	66	20	12	6	18
Uganda	1967	66	32	16	11	21
Zambia	1965-1966	96	92	88	41	62

Source: Richard Jolly and Christopher Colclough, "African Manpower Plans: An Evaluation," *International Labor Review* 106, August/September 1971, p. 210.

[a] A = university degree or equivalent; B = higher secondary school examination generally taken after six or seven years of secondary schooling; C = form II plus two to four years of formal training; and D = form II or primary school plus one to four years of formal training.

[b] Includes all non-Africans for 1964 and Western expatriates only in 1969.

enrolled in school. The enrollment rate for ages 6-23 was only 17.5 percent for the subcontinent. About half as many Africans attended school as in other developing areas, and Africans generally stayed in school for a shorter time than their counterparts elsewhere.[24]

Not surprisingly, almost all of the newly independent African governments placed a high priority on education. Per capita expenditures on education grew from about 60 percent of the average for developing countries in 1965 to 110 percent by 1980.[25] By 1985 African enrollment rates for children ages 6-11, 12-17, and 18-23 had increased to 55, 49, and 8 percent, respectively, considerably closer to developing country averages than they had been in 1960.[26]

Aid donors played a significant role in extending the relatively few higher education opportunities to Africans. The number of students financed by DAC countries to obtain advanced degrees rose from negligible levels in the early 1960s to nearly 15,000 in 1970. Asia received assistance for only 1,500 more trainees than Africa in the same year.[27] DAC countries provided 32,600 study and training grants to Africans in 1979 (more than twice as many as in 1970), versus about 31,000 to Asians.[28] Table 2.2 illustrates the total number of higher-level students studying abroad, not necessarily financed by DAC countries, for selected regions and years. The table indicates that substantial numbers of Africans, both absolutely and as compared with other regions, have received university training abroad.

TABLE 2.2

HIGHER-LEVEL STUDENTS STUDYING ABROAD

Region	1968	1977	1987
Africa (including North Africa)	45,497	126,279	172,733
South America	23,424	47,553	34,313
Asia	174,166	291,828	460,035

Source: UNESCO, *Statistical Yearbook*, various years.

To a considerable extent, then, technical cooperation in the decades after independence meant opening the educational systems of the developed world more widely to African students. But there was also a substantial movement of technical assistance personnel to African countries, to fill gaps throughout the public sector and to teach in burgeoning educational systems.

Table 2.3 illustrates the geographic distribution of technical assistance personnel from OECD countries serving overseas in 1963. Of the approximately 44,000 nonteacher technical assistance personnel serving worldwide in 1963, nearly half served in Sub-Saharan Africa, which claimed four times as many nonteacher technical assistance personnel in 1963 as all of Asia. Of the approximately 31,000 technical assistance personnel serving in Sub-Saharan Africa in 1963, about 10,000 served as teachers.

Throughout the 1960s, technical cooperation resources claimed an expanding share of an expanding development assistance pie. By 1970, technical cooperation claimed 48 percent of bilateral overseas development assistance directed to Sub-Saharan Africa in contrast with 27 percent of overseas development assistance worldwide. Multilateral technical cooperation grew from negligible levels in 1960 to $93 million in 1970 (about 25 percent of total multilateral resources in both years).[29]

By 1970, the total number of technical assistance personnel from OECD countries serving in Sub-Saharan Africa had expanded to nearly 42,500 people. The number of expatriates in education increased by more than 50 percent to more than 16,000 and the number in nonteaching fields increased by approximately 25 percent to more than 26,000.[30] The composition of experts changed as well. In 1963, France and the United Kingdom provided more than 75 percent of technical assistance personnel to Sub-Saharan Africa. During the 1960s, growth in technical cooperation resources came primarily from noncolonial bilateral and multilateral sources, particularly the United States and the United Nations. But in the 1960s and 1970s, as today, the technical assistance community in most African countries remained dominated by nationals of former colonial powers.

Although technical cooperation retained throughout the 1970s its positive popular image and its strong support among both donors and

TABLE 2.3

GEOGRAPHIC DISTRIBUTION OF TECHNICAL ASSISTANCE PERSONNEL FROM OECD COUNTRIES SERVING OVERSEAS IN 1963

Country/Region	Total	Of Which Teachers	Belgium	France	U.K.	U.S.	Other OECD
America	4,609	[1,019]	-	597	580	3,268	164
Africa	66,623	[33,366]	2,660	48,535	10,380	3,346	1,602
[Sub-Saharan Africa]	[31,262]	[10,122]	[2,655]	[13,888]	[10,317]	[3,019]	[1,383]
Asia	8,640	[3,176]	-	1,686	1,752	4,327	875
Other	2,156	[685]	-	379	836	831	470
Total	82,028	[38,246]	2,660	51,197	13,548	11,512	3,111

Source: Angus Maddison, "Foreign Skills and Technical Assistance in Economic Development," Development Centre, Organization for Economic Co-operation and Development, 1965, p. 21.

recipients, some observers were raising criticisms that were to find much broader expression in the late 1980s: that recruitment of good people was difficult, that job descriptions were imprecise, that much work actually performed by technical assistance personnel was not in accord with job descriptions, and that a manpower-scarce environment meant rapid job turnover of counterparts and limited training impact.[31]

Some negative implications of colonial traditions for capacity-building or institution-building technical cooperation also became evident in the postcolonial period. The tradition of expatriate staff filling line positions was deep rooted, and in the circumstances of the time, this had some effects that obstructed effective training on the job, technology transfer, and institution building.

A study of French technical assistance in Senegal in the early 1970s provides an interesting case in point.[32] The author cites the "absence of explicitly stated objectives" as one of two main contributing factors to the "disappointment" with technical assistance. Some of the study's other findings illustrate potential problems with the technical assistance efforts of the 1960s in terms of capacity building. Sixty percent of French technical assistance personnel in Senegal had been in the colonial service. Two-thirds of all technical assistance personnel interviewed in the study saw "no difference between the jobs they did in France and those they were doing in Senegal." The majority of those surveyed were opposed to accelerated Africanization of administrations.

The slow dissipation of this colonial heritage is only one of many factors explaining the large and persistent presence of technical assistance in Sub-Saharan Africa. Others, some of which are undoubtedly more important, are considered in later chapters. What stands out in this brief historical sketch is the heavy weight of the colonial situation in explaining the continent's extensive reliance on technical assistance in the years after 1960.

QUANTITATIVE TRENDS

In this section we address several questions. What are recent trends in technical cooperation provision and what is the present position with respect to the volume and nature of this form of aid? Who are the main providers and the main recipients? What types of technical cooperation are being provided? This discussion draws on the extensive data set out in the Statistical Annex.

Data Uncertainties

The major source of data on technical cooperation is DAC, which has kept statistics on development cooperation for over two decades. DAC data understate the true volume of technical cooperation expenditures because the data include only free-standing technical cooperation financed by grants from OECD member countries and from the official multilateral institutions. They thus do not include four important categories of technical cooperation:
- Investment-related technical cooperation (10-20 percent of the total, according to UNDP estimates);
- Technical cooperation from non-OECD member countries (for example, ex-Soviet Union, Cuba, Egypt, Algeria, and China). That this omission can be significant is indicated by experience in Guinea, where in the late 1980s 13 percent of all technical cooperation resources were provided by non-OECD-member bilaterals);
- Technical cooperation financed by nongovernmental organizations, which, according to some UNDP estimates based on country data, could amount to as much as 10 percent of total technical cooperation flows. (The World Bank's 1991 report of the Technical Assistance Review Task Force made a similar estimate, based on aggregate data); and
- Technical cooperation that is loan financed, of which the World Bank is the largest provider: it financed about $350 million of such technical cooperation annually in Africa in the late 1980s.

Trends in Technical Cooperation Grants to Sub-Saharan Africa (1970-1989)

In 1970, Sub-Saharan Africa received $435 million in technical cooperation grants. Since then, the volume in nominal terms has risen dramatically—by about 800 percent. Since the early 1980s, the nominal rise has been less dramatic but still substantial, climbing from around $2 billion in 1980 to over $3 billion in 1989. In real terms, however, this increase is not nearly as strong. Between 1975 and 1980, real technical cooperation grants grew by almost 50 percent, from $2.2 billion to $3.1 billion. And, as Table 2.4 and Figure 1 show, technical cooperation grants were generally stagnant in the 1980s, fluctuating slightly around 3.2 billion 1988 dollars. Technical cooperation grants as a proportion of official development assistance have decreased over the last two decades (Figure 2). In the early 1970s, technical cooperation grants fluctuated around 40 percent of total official development assistance. By the late 1980s, this figure had dropped to a level of 24 percent.[33]

How Many Technical Cooperation Personnel?

The number of technical assistance personnel that these financial flows finance is not well established. In the early 1970s, OECD reported 42,500 technical assistance personnel working in Sub-Saharan Africa. Since then, OECD has faced great difficulties in compiling comparable data. There are too many gaps in the data provided by donors to discern a trend.[34]

It often happens that when important numbers are lacking, intrepid writers more or less invent them. These numbers are then snapped up and cited everywhere. Something of this kind happened with the total number of technical assistance personnel in Sub-Saharan Africa. The figure of 80,000 technical assistance personnel in Africa in the 1980s is widely quoted. But it is hard to track down how that number was derived. It seems to have appeared first in an internal World Bank report in the early 1980s but was a rough estimate, apparently based on no systematic survey.

Recent surveys of technical cooperation personnel carried out in 12 countries strongly suggest that this figure of 80,000 is substantially overestimated. Many of the 45 Sub-Saharan African countries report total resident personnel at under 500 persons, including volunteers and personnel financed by nongovernmental organizations. Of the 22 countries with recent data listed in Table 2.5, only 11 had 500 or more technical assistance personnel. These findings imply a total of resident personnel closer to 40,000 than to 80,000. A figure much below 80,000 in Sub-Saharan Africa is also consistent with data on trends in this

TABLE 2.4
TRENDS IN TECHNICAL COOPERATION GRANTS TO SUB-SAHARAN AFRICA, REAL AND NOMINAL
(billions of 1988 dollars)

Real Values of Technical Cooperation Grants and Official Development Assistance and Deflator for OECD Countries

ODA/TCG	1975	1976	1977	1978	1979	1980	1981	1982	1983	1984	1985	1986	1987	1988	1989
Nominal ODA	3.1	3.2	3.6	4.9	6.2	7.4	7.4	7.5	7.3	7.6	8.6	10.6	11.9	13.2	13.7
Real ODA	6.8	7.1	7.3	8.4	9.5	10.5	10.8	11.3	11.0	11.7	13.1	12.7	12.8	13.2	13.9
Nominal TCG	1.0	1.0	1.0	1.4	1.7	2.2	2.2	2.0	2.0	2.0	2.2	2.7	3.0	3.2	3.2
Real TCG	2.2	2.2	2.2	2.3	2.6	3.1	3.2	3.1	3.1	3.1	3.4	3.3	3.2	3.2	3.3
Deflator[a]	44	45	49	58	64	70	68	66	66	65	66	83	93	100	99
Indexed (1975=100)															
Nominal ODA	100	107	120	162	203	245	243	248	241	251	285	350	393	437	454
Real ODA	100	104	107	123	140	154	159	166	162	172	193	187	188	194	204
Nominal TCG	100	100	107	136	166	217	217	202	203	199	220	273	295	323	321
Real TCG	100	97	96	103	115	137	142	135	136	136	149	146	141	143	144

Source: Statistical Annex.
[a] Deflator is taken from the 1990 "Development Cooperation Report," UNDP, p. 267.

FIGURE 1 TRENDS IN TECHNICAL COOPERATION GRANTS

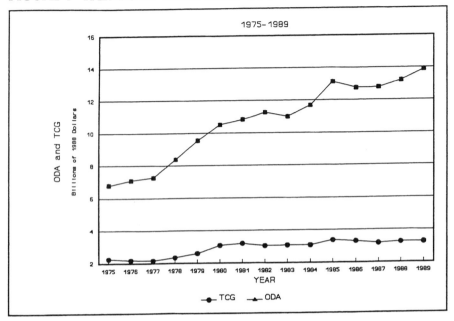

FIGURE 2 TECHNICAL COOPERATION GRANTS AS A PART OF OFFICIAL DEVELOPMENT ASSISTANCE

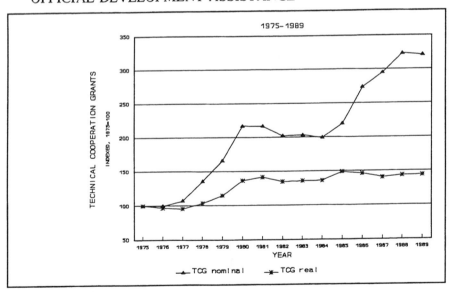

TABLE 2.5

TECHNICAL COOPERATION PERSONNEL BY CATEGORY: SELECTED COUNTRIES, 1988 (a) or 1989 (b)

	Experts		Volunteers		
Countries 1989	Int'l	Nat'l	Int'l	Nat'l	Total
Benin (a)	291	0	149	5	445
Burkina Faso (a)	441	22	195	20	678
Burundi (a)	596	88	276	0	960
Cape Verde (a)	198	1	36	0	179
Central African Republic (a)	520		237		757
Chad (b)	290	12	72	0	374
Comoros (b)	161	1	132	1	295
Gabon (a)	877	-	-	-	-
Gambia (a)	82	3	32	0	117
Guinea (a)	638	5	67	0	710
Guinea Bissau (a)	-	-	-	-	572
Guinea Bissau (b)	363		153		516
Equat. Guinea (b)	-		-	-	300
Madagascar (b)	739	12	108	71	930
Mali (b)	804	45	327	0	1176
Mozambique (a)	-	-	-	-	2290
Rwanda (a)	412	187	345	16	960
Sao Tome & Principe (b)	-	-	-	-	122
Senegal (a)	1573	-	-	-	-
Swaziland (a)	221	3	108	0	332
Tanzania (a)	1112	83	403	0	1598
Togo (a)	241	17	178	0	436

Sources: UNDP "Development Cooperation Reports" and government data, 1989 and 1990.

area of some leading bilateral donors. The number of French technical assistants (*coopérants*), for example, has fallen from 9,519 in 1984 to 6,208 in 1991, a drop of some 35 percent in the last eight years.[35] The total of United Kingdom personnel working worldwide declined from 11,500 in 1972 to 4,900 in 1980 and 4,100 in 1988. Excluding volunteers, the number dropped from 4,200 to 2,200 between 1980 and 1988.[36] In Africa, the total number of person-years (including volunteers) of technical assistance personnel provided to Sub-Saharan Africa by the United Kingdom fell from 2,399 in 1982 to 1,706 in 1988—or by almost 30 percent.[37]

The data for financial costs raise further doubts about the validity of the 80,000 resident personnel estimate. If we assume an average cost for person-year of personnel of $100,000, and assume also that 75 percent of the $3.2 billion spent on technical cooperation in 1989 is for personnel, there would have been only about 23,000 personnel resident in 1989. Not all technical assistance personnel are long term and resident, of course, and average costs may be less than $100,000 per year. But it is most unlikely that these differences could account for much more than 30,000 technical assistance people.

According to country surveys carried out since 1988 under NaTCAP programmes, the number of long-term technical assistance personnel in Sub-Saharan African countries, including volunteers, usually varies between 350 to 1,000 persons, with the fewest (117) in Gambia, and the most (2,290) in Mozambique.

The significance of these technical assistance personnel in terms of enhanced capacity to manage development varies greatly from country to country, depending on the size of the civil service at the higher professional levels and the level of human resource development. In several countries, expatriate personnel represent a significant portion of the high-level personnel in the public sector. In Guinea-Bissau, there were about 500 expatriate technical assistance personnel in 1987, 20 percent more than the 400 university-trained nationals in government employment. This situation was particularly marked in key Ministries such as Plan, Finance, and Health.

In Chad, the ratio of expatriate technical assistance personnel to government staff with university education is 1:5. In some Ministries such as Health, Public Works, and Agriculture, the ratio is 1:3. In Burundi, in the health sector, expatriate personnel account for about a third of all doctors, and in rural areas these technical assistance personnel make up most of available physicians. In Senegal, a country renowned for its human resources, the ratio is still 1:10.

Skill Levels and Sectoral Composition

The composition of technical assistance personnel is shown in Tables 2.5 and 2.6. The following points stand out.

- National personnel still constitute an insignificant element of technical assistance personnel.
- There is a surprisingly high percentage of personnel with qualifications at the level of skilled worker or technician (two years of postsecondary education) as opposed to graduate degrees—around 15 percent in Togo, Benin, and Mozambique, and around 10 percent in Burkina, Cape Verde, Guinea, and Tanzania. These would seem to be easy to replace, though the type of practical expertise they possess is often critical to the success of development undertakings.
- Volunteers are also major players among the technical assistance personnel, accounting for 20 to 40 percent on average of the total. The important role of volunteers is masked in statistics on technical cooperation and technical assistance expenditures, because the cost of a volunteer is a fraction of that of full-scale experts.
- Many of the technical assistance personnel continue to come from industrialized countries, largely because most donors recruit from among their own nationals. However, in the U.N. system, this trend has been reversed: of the 18,000 posts of experts and consultants financed worldwide by UNDP in 1988, 10,000 were filled by developing country nationals.
- The sectoral distribution of technical assistance personnel follows a markedly similar pattern in recipient countries: education, health, and rural development are the three major users, accounting together for two-thirds or three quarters of total num-

TABLE 2.6
SPECIALIZATION OF TECHNICAL ASSISTANCE PERSONNEL, SELECTED COUNTRIES, 1989
(percentage of the total)

Country	Unclassified	Sector									Unclassified		Total[a]
		Administration	Agriculture	Communication	Education	Health	Resettlement	Engineers	Technical	Skilled Trades	Natural Sciences	Social Sciences	
Benin	2	6	25	1	8	23	0.5	16	13	1	0.7	2	445
Burkna Faso	-	10	20	5	29	10	2	8	8	1	0.6	8	678
Burundi	18.5	7	15	2	17	16	2	13	4	1	1	4	960
Cape Verde	2	5	9	2	19	19	2	19	5	5	0	13	179
CAR	-	-	16	-	40	16	-	-	-	-	-	-	757
Chad	18	7	23	3	18	18	-	-	-	-	-	-	374
Comoros	12	6	8	4	48	16	-	-	-	-	-	-	295
Eq. Guinea	38	10	-	-	38	29	-	-	-	-	-	-	300
Gabon	20	6	2	4	63	12	0.3	5	2	0.1	2	1	-
Gambia	4	21	15	0	18	19	1	7	1	2	1	11	117
Guinea	-	11	15	0.5	16	4	1	32	9	1	5	6	710
Guinea B	14	12	10	1	14	29	-	-	-	-	-	-	516
Madagascar	24	3	14	1	24	23	-	-	-	-	-	-	930
Mali	22	9	29	2	22	10	-	-	-	-	-	-	1176
Mozambique	-	13	11	9	12	15	7	9	16	-	-	-	2290
Rwanda	2	13	15	1	8	20	2	9	7	0.5	4	15	960
Sao Tome	44	7	-	-	44	39	-	-	-	-	-	-	122
Senegal	4	5	15	1	48	11	0.5	-	-	-	-	-	-
Swaziland	0	14	11	2	45	13	2	6	0.6	0.4	2	5	332
Tanzania	8	18	11	2	7	10	4	20	4	7	3	4	1598
Togo	0	5	26	1	33	9	0.6	5	10	6	0	5	436

Sources: Government data and "Development Cooperation Reports," UNDP, 1989 and 1990.
[a] Absolute number.

bers of technical assistance personnel. These include large numbers of teachers, doctors, and paramedical personnel, mostly performing substitution functions. This situation may be particular to the 13 countries for which data are available, because only one is a former British colony while others are former French, Portuguese, and Spanish colonies. The historical tradition of this type of assistance in the social sectors is less strong in the Anglophone countries, and there may well be more technical assistance personnel in the other productive sectors in these anglophone countries.

Sources and Recipients of Technical Cooperation: Patterns of Concentration

Table 2.7, drawn from Annex Table 3, shows trends over the past 20 years in sources of technical cooperation. Five main tendencies are noteworthy.

- Bilateral technical cooperation continues to dominate, although the multilateral share rose by a third over the decade of the 1980s. This finding understates the multilateral share because it excludes loan-financed technical assistance. For the World Bank alone, this amounted to over $350 million annually in the late 1980s.
- Bilaterals finance about three quarters of technical cooperation grants, and multilaterals about one quarter. Because these data do not show investment-related technical assistance, financing from bilateral donors (such as the United States) that do not have special technical assistance or technical cooperation programmes may be understated.
- In 1989, five major bilaterals—France, Germany, the United States, the United Kingdom, and the Netherlands—accounted for over 50 percent of all technical cooperation financing, and two-thirds of bilateral technical cooperation. With the UNDP, these are the main providers of technical cooperation to Africa.
- The former colonial powers, France, the United Kingdom, and Belgium, still play a major role, but their importance has declined from 44 to 26 percent of all technical cooperation grants.

TABLE 2.7

TRENDS IN TECHNICAL COOPERATION GRANTS
DISBURSEMENTS FROM MAJOR DONOR SOURCES, 1970-1989
(current $US millions and percentage)

Donor	1970 $ million	1970 Percentage	1980 $million	1980 Percentage	1985 $ million	1985 Percentage	1989 $ million	1989 Percentage
France	110.2	25.3	510.8	23.5	341.4	15.6	531.9	16.5
Germany	37.7	8.7	307.9	14.1	246.8	11.2	410.6	12.7
USA	47.0	10.8	160.0	7.4	236.0	10.8	341.0	10.6
UK	39.5	9.1	193.5	8.9	104.0	4.7	197.9	6.1
Netherlands	8.4	1.9	101.6	4.7	72.5	3.3	178.5	5.5
Belgium	43.3	9.9	170.4	7.8	80.9	3.7	106.7	3.3
Japan	2.4	0.6	32.0	1.5	45.1	2.1	114.6	3.6
Italy	6.5	1.5	16.8	0.8	157.5	7.2	119.8	3.7
Sweden	8.4	1.9	46.8	2.3	55.1	2.5	115.4	3.6
Other bilat.	38.6	8.9	143.7	6.6	176.6	8.0	196.5	6.1
Total bilat.	342.0	78.5	1683.5	77.4	155.9	69.1	2312.9	71.7
Multilateral[a]	93.5	21.5	492.6	22.6	679.3	30.9	908.1	28.2
Total	435.5	100.0	2176.1	100.0	2195.2	100.0	3224.3	100.0

Source: OECD, *Geographic Distribution of Financial Flows to Developing Countries*, Paris, various years.
[a] Underestimated for reasons noted in the text. For example, technical cooperation financed by loans is excluded; in the late 1980s, the sum of technical cooperation loans to Sub-Saharan Africa was over $350 million annually. World Bank, "Managing Technical Assistance in the 1990's: Report of the Technical Assistance Review Task Force," 1991.

The relative importance of Germany, the Netherlands, Italy, Sweden, and Japan has steadily increased over this same period.

- Most of the technical cooperation from the former colonial powers goes to their former colonies. The proportion is about 85 percent for France, with the Côte d'Ivoire receiving almost

15 percent of all French technical cooperation to Sub-Saharan Africa. Some 80 percent of British technical cooperation goes to the Commonwealth Sub-Saharan African countries, with 35 percent of this total going to Kenya and Uganda. About 80 percent of Belgian technical cooperation goes to Zaire, Rwanda, and Burundi, and 80 percent of Italian-financed technical cooperation goes to Ethiopia and Somalia. Of Portugal's technical cooperation, 90 percent goes to African Lusophone countries.

These patterns of concentration are also evident from the recipient side; the former colonial powers represent the main source of technical cooperation assistance received by these countries. Six Sub-Saharan African countries, for example, receive more than one-half of their technical cooperation resources from France: Central African Republic, Congo, Côte d'Ivoire, Djibouti, Gabon, and Madagascar. To a slightly lesser extent, Belgium was the dominant technical cooperation donor in Zaire as was the United States in Liberia. However, 14 countries, all in the U.N.'s "least developed" category, receive more than one-third—and two countries receive more than one-half—of their technical cooperation assistance from multilateral donors, mainly UNDP.[38]

The other major bilateral donors, Germany, United States, Netherlands, Japan, and Sweden, as well as the multilaterals have spread their technical cooperation more evenly. However, some donors such as Sweden, the Netherlands, and UNDP tend to favor the least developed countries.

It is apparent that the volume of technical cooperation resources a country receives depends on a variety of factors, heavily weighted by the colonial past. There is little connection between technical cooperation disbursements and the level of development and relative need for technical cooperation. The least developed countries appear to receive more than the rest of Sub-Saharan Africa; their average technical cooperation receipts per capita in 1989 were $10 compared with $7 for Sub-Saharan Africa as a whole. Six countries—Ethiopia, Tanzania, Sudan, Kenya, Zaire, and Senegal—received almost a third of all technical cooperation resources allocated for Africa in 1988. These are all large countries in population. The five countries that received most technical cooperation per capita all have small populations, with the exception of Madagascar. The five countries for which technical cooperation represented the largest proportion of GNP have small GNPs. This is illustrated in Table 2.8.

Technical Cooperation as a Macroeconomic Variable: Its Weight in Recipient Economies

The economic significance of technical cooperation is far from negligible in much of Africa. Technical cooperation provides a substantial share of government revenues and expenditures,[39] of development spending, and of total skilled human resources in some countries. Until very recently, recipient countries have not kept reliable quantitative records on such matters as donor expenditures on technical cooperation, related government costs, and numbers and types of personnel and training programmes involved. A growing number of Sub-Saharan African countries are correcting this situation by setting up improved data systems through the NaTCAP process. Since 1988, many countries have carried out baseline surveys of the technical cooperation situation and maintain them annually. Review of these country data as well as more aggregate OECD statistics provides some indications of overall trends.

Although the importance of technical cooperation varies a great deal by country, it is trivial in very few. Whatever its importance, it has, unfortunately, been a consistently neglected resource. In relation to its financial and developmental significance, technical cooperation has been undermanaged and underutilized by recipient governments (see Chapter Four).

Technical Cooperation as a Financial Resource

On average, technical cooperation resources represent about one quarter of all external aid, but its share varies among countries, from 10 to 60 percent. For 13 countries, technical cooperation in the late 1980s represented over a third of net official development assistance. These statistics are based on OECD data that, as noted above, probab-

TABLE 2.8

MAJOR TECHNICAL COOPERATION RECIPIENTS, 1989

Total Expenditures		TC Per Capita		TC as Percentage of GNP	
Country	$US mill.	Country	$US	Country	$US
Ethiopia	218	Seychelles	134	Guinea B.	14.5
Tanzania	199	Djibouti	83	Gambia	12.0
Sudan	183	Madagascar	56	Somalia	11.3
Kenya	174	Botswana	47	Comoros	8.0
Zaire	150	Cape Verde	46	Mozambique	7.4

Source: Statistical Annex.

ly underestimate true technical cooperation expenditures because investment-related programmes, activities financed by non-governmental organizations, loan-financed technical cooperation, and technical cooperation from non-OECD sources are not included.

The significance of technical cooperation as a public resource is apparent when it is compared with the government's own revenues. In 1989, technical cooperation was equal in size to 14 percent of government revenues in Sub-Saharan African countries excluding Nigeria. For 10 countries, it was equal to at least 30 percent. Technical cooperation is also a significant source of foreign exchange and should be evaluated in the light of the difficulty many Sub-Saharan African countries have had in generating such resources. In 1988, technical cooperation expenditures equaled 15 percent of the value of exports of

Sub-Saharan Africa (excluding Nigeria). For 18 countries, this was over 30 percent.

Although the real value of technical cooperation expenditures has grown only by around 5 percent over the last decade, its importance in the country context has grown substantially. Partly because of economic stagnation or deterioration in so many Sub-Saharan African countries in the 1980s, the average share of technical cooperation as compared with government revenues and with export earnings has doubled during the 1980s. Moreover the statistics on technical cooperation inflows reviewed here represent a significant underestimate of total technical cooperation.

Input Composition of Technical Cooperation Expenditures

Somewhat surprisingly, given the volume of debate in recent years about changing the technical cooperation input mix in favor of training and supplies and equipment, the personnel component remains predominant. Data from country surveys carried out for NaTCAP reviews summarized in Table 2.9 show that, for 8 of the 10 countries for which recent data exist, the personnel component represented more than 60 percent of all financing allocated to technical cooperation programmes received by the country.

Financing of equipment and operating costs are generally limited to facilitating the work of experts. Items covered are vehicles, office equipment, and some minor operating materials such as seeds or chemicals for agricultural research. One frequently heard complaint of recipient government officials is that not enough funds are allocated for the financing of equipment that might be left after the technical assistance people leave. A survey carried out in Chad, for example, revealed the perception that this underfinancing of equipment is considered a major constraint to sustainability of projects.[40]

Training programmes are also a fairly minor part of technical cooperation programmes. For the 10 countries referred to above, they represent 10 to 15 percent of all technical cooperation expenditures. However, this does not include expenditures for teachers or on-the-job training provided by experts to national counterparts. Recipient coun-

TABLE 2.9

TYPE OF TECHNICAL COOPERATION EXPENDITURES, SELECTED COUNTRIES, c. 1989
(percentage)

Country	Personnel	Training	Equipment and Management
Benin	80.5	15.0	4.5
Cape Verde	72.9	16.4	10.7
CAR	85.6	11.5	2.9
Chad	52.8	9.1	38.2
Comoros	85.6	11.5	2.9
Gambia	57.8	9.9	32.3
Guinea	78.8	11.8	9.4
Madagascar	86.0	12.8	1.2
Mali	82.2	13.2	4.6
Senegal	77.2	9.3	13.5
Average	75.9	12.1	12.0

Source: UNDP "Development Cooperation Report," various years.

try spokesmen often express requests for more training. For example, the 1991-1993 technical cooperation programmes for four priority sectors in Burundi all emphasize the need for increased training programmes. This is in striking contrast to other inputs for which little increase was asked.

Summing Up

The trends in technical assistance personnel, then, show no dramatic shifts over the past 10 years. The available aggregate data on numbers of technical assistance personnel in Sub-Saharan Africa do not reflect the cuts that some donor statistical sources show for the 1980s. Instead, the aggregates are stagnant, or even show an increase in the size of technical cooperation programmes, as measured by the volume of financial flows.

The composition of technical cooperation programmes had not changed much either, by the end of the 1980s. Most of the financing is for long-term, expatriate personnel. And a surprisingly large proportion of these personnel carry out operational, direct-support types of activities. A significant proportion is middle-level technical personnel, such as agricultural technicians, paramedics, mechanics, and accountants. Such personnel are often critical to successful development efforts, and seem to be harder to replace than formal educational qualifications would suggest.

The continuing presence of many technical assistance personnel was not expected to be so enduring a phenomenon. Nor was the persistent presence of large numbers of long-term resident personnel anticipated. In 1962, President Senghor said, in reviewing training programmes for Senegalese nationals: "Grâce à cette méthode réaliste et progressive nous sommes persuadés que dans dix ou vingt ans d'ici nous n'aurons plus besoin de techniciens étrangers."[41] Thirty years later the number of technical assistance personnel resident in Senegal had not significantly changed.

This situation may to some extent be explained by a continuing high level of genuine demand related to the introduction of new technology (such as computers or new farming system research), and the growth of newly implanted institutions, universities, and research institutes, for example. But the scale and composition of technical assistance personnel cannot be justified on this count, nor is this enough to explain the persistence of large volumes of such personnel into the 1990s. Much of this use of technical assistance personnel seems out of harmony with the expressed priorities of African countries, which call for such personnel only in highly specialized technical areas and for specialized training of national personnel who have the requisite basic education. It is also inconsistent with the professed policy of many recipient governments, and of many donors, which define creation of sustainable capacity as the first task of technical cooperation.

NOTES

1. The French government does not have a term to cover the general notion of technical cooperation or technical assistance. The term *assistance technique* refers to one specific programme: the provision of expatriate personnel financed by the Ministry of Cooperation under two-year renewable contracts; the people thus provided are referred to as *coopérants* or *assistants techniques*. *Assistance technique* so defined does not include other types of personnel or resources for technology transfer, such as personnel financed by institutions other than the Ministry of Cooperation, short-term consultants or long-term personnel in projects (*experts*), or volunteers (Volontaires du Service National—VSN). Because France is the predominant donor in Francophone Africa, this definition of the term *assistance technique* to mean provision of long-term expatriate personnel has been widely adopted in those countries.

2. According to DAC, technical assistance refers exclusively to "inputs of labor or expertise provided as part of a capital project" whereas technical cooperation refers to development assistance activities, including technical assistance personnel and training, that are not attached to capital assistance projects but are aimed primarily at increasing "the stock of (the recipient's) human intellectual capital." DAC, like the World Bank, calls this latter activity "free-standing." OECD, DAC, *Geographical Distribution of Financial Flows to Developing Countries, 1986-1989*, Paris, 1991, p. 339.

3. This was the rationale behind a UNDP policy decision, taken in 1970, to define technical cooperation programmes by the objectives to be achieved and to introduce considerable flexibility with regard to the range of inputs to be financed.

4. Marc de Bernis, "La coopération technique—de quoi parle-t-on?" Internal UNDP paper, September 1989.

5. Herein lies one reason technical cooperation has become so closely identified with inputs. Where technical cooperation projects are not specified as such, it is difficult to identify projects with primarily technical cooperation objectives but much easier to identify inputs. In these cases, technical cooperation becomes synonymous with personnel.

6. In French, the *coopérant* arrangement is referred to as *assistance technique de substitution, mise en disposition*, or *assistance directe*; the latter is called *(assistance) dispensée dans le cadre des projets de développement*. See Witold Mikulowski, "Les effets réels de la coopération administrative: Quel jugement porter sur le terrain?" In *Revue Française d'Administration Publique*, avril-juin, 1989, no. 50, p. 300.

7. See, on classifications of technical assistance, George Honadle, David Gow, and Jerry Silverman, "TA Alternatives for Rural Development: Beyond the Bypass Model," *Canadian Journal of Development Studies*, vol. IV, no. 2, 1983; and David Gow, "Provision of Technical Assistance: A View from the Trenches," *Canadian Journal of Development Studies*, vol. IX, no. 1, 1988.

8. Organization for Economic Co-operation and Development, Development Assistance Committee, "Principles for New Orientations in Technical Cooperation," Paris, 1991.

9. World Bank, "The African Capacity Building Initiative: Toward Improved Policy Analysis and Development Management in Sub-Saharan Africa," Washington, D.C., 1991.

10. World Bank, *Sub-Saharan Africa: From Crisis to Sustainable Growth. A Long-Term Perspective Study,* vol. I, 1989, p. 190.

11. This analysis draws on Milton Esman, "Strategies and Strategic Choices for Institutional Development," paper prepared for Conference on Institutional Development and the World Bank, December 14-15, 1989; and Coralie Bryant, "Development Management and Institutional Development: Implications of their Relationship," American University and Overseas Development Council, Washington, D.C., October 1985.

12. This definition is based on Edgar Schein, *Organizational Psychology*, Second Edition, New York: Prentice Hall, 1979.

13. Arturo Israel, *Institutional Development—Incentives to Performance*, Johns Hopkins University Press, 1987, p. 1. An important internal study of the World Bank defines institutional development as: "The creation or reinforcement of the capacity of an organization to generate, allocate and use human and financial resources effectively to attain development objectives, public or private." (Beatrice Buyck, "Technical Assistance as a Delivery Mechanism for Institutional Development: A Review of Issues and Lessons of Bank Experience," World Bank, 1989, p. 4.

14. See Esman, 1989, pp. 25, 41.

15. See, for example, footnote 13 for the definition proposed by Buyck, 1989.

16. It should be noted that this definition is much broader than that given in the World Bank's recent (1991) internal review, "Managing Technical Assistance: Report of the Technical Assistance Review Task Force." This document defines technical assistance or technical cooperation as "the transfer or adaptation of ideas, knowledge, practices, technologies or skills in pursuit of development." It then distinguishes four types of technical assistance: (1) policy development, which helps decision makers in macroeconomic policy or sector management—for example, studies showing trade-offs among policy options, and seminars; (2) institutional development technical assistance, which is defined as "improving the ability of one or more government agencies (or private entities) or reforming the regulatory/procedural framework. It is normally specific to an entity, sector or program"; (3) capacity building, which is defined as "enhancing domestic capacity, usually multisectoral or function-specific (for example, accounting, information technology, or research capacity); it is not limited to a particular entity or program; and (4) project or program support, for "facilitating one or more stages of the project/program cycle" (p. 42).

17. The World Bank, *Accelerated Development in Sub-Saharan Africa: An Agenda for Action,* 1981.

18. D.K. Fieldhouse, *Black Africa 1945-1980: Economic Decolonization and Arrested Development,* London: Allen and Unwin, 1983, p. 35.

19. United Nations, *Economic Survey of Africa Since 1950,* Department of Economic and Social Affairs, 1959, p. 47.

20. Rita Cruise O'Brien, "Colonization to Co-operation? French Technical Assistance in Senegal," *Journal of Development Studies*, October 1971, pp. 45-57.

21. "L'assistance technique aux pays insuffisamment développés: Première partie," *La Documentation Française*, 1954, p. 4.

22. Ibid., Troisième partie, p. 11. Many more scholarships and other forms of assistance were provided through the Ministère des Colonies.

23. United Nations, "15 Years and 150,000 Skills: An Anniversary Review of the United Nations Expanded Programme of Technical Assistance," Technical Assistance Board, 1965, pp. 13 and 23.

24. For developing countries as a whole, the numbers for the same age groups were 48 percent, 35 percent, and 7.5 percent, respectively. The total enrollment ratio for developing countries as a whole was 33 percent.

25. This superior relative performance could not be maintained, given slow growth and other difficulties in the region. By 1987, per capita education expenditures in Africa were only 52 percent of developing country average expenditure (UNESCO, *Statistical Yearbook*, various years).

26. The averages for all developing countries for the same age groups were 74, 44, and 13.

27. Organization for Economic Co-operation and Development, "Development Assistance: Efforts and Policies of the Members of the Development Assistance Committee," 1971.

28. Ibid., 1982, p. 113.

29. OECD, DAC, *Geographical Distribution* . . . , 1991, various years.

30. OECD, 1971 p. 170.

31. Denyse Harari, "The Role of the Technical Assistance Expert," OECD Development Center, 1974; and Milton J. Esman and John D. Montgomery, "Systems Approaches to Technical Cooperation: The Role of Development Administration," *Public Administration Review*, September/October 1969.

32. O'Brien, 1971, p. 47-48.

33. Although total real official development assistance worldwide declined by 3 percent from 1980 to 1989, total real ODA to Sub-Saharan Africa rose consistently. Total real technical cooperation grants to Sub-Saharan Africa also rose, but at a slower rate than total real ODA to Sub-Saharan Africa.

34. Information from the DAC secretariat. It should be noted that a large number of technical assistance personnel come from non-OECD countries.

35. Most French technical assistance goes to Sub-Saharan Africa. The biggest fall was in teachers, whose number declined by 24 percent between 1984 and 1989—from 6,823 to 4,924. In 1991, there was another 15 percent decline—to about 4,300. The number of nonteachers ("technicians") fell by much less—from 2,696 in 1984 to 2,423 in 1989 (-10 percent) and to about 2,000 in 1991. Almost 85 percent of French technical assistance (5,312 persons) is in 10 countries: Côte d'Ivoire (1,351); Senegal (761); Gabon (511); Cameroon (413); Mali (343); Niger (291); Central Africa Republic (286); Congo (269); Burkina (259); and Mauritania (245). Data were provided by the French Ministry of Cooperation.

36. G. Glentworth, "Technical Cooperation in the UK Aid Programme," paper given at Joint EDI/KIA Seminar on the Management of Technical Assistance, Nairobi, April 17-27, 1990.

37. Overseas Development Administration, *British Aid Statistics, 1980-1984* and *1985-1988*, Government Statistical Service, 1985 and 1990. Belgian-financed technical assistance personnel have also declined overseas, from 1,244 in 1987 to 936 in 1989. Volunteers (associate experts) declined in the same period from 224 to 188. Because most Belgian personnel were in Zaire, their number has undoubtedly declined sharply since 1989. (Abog, Aktiviteiten Verslag, 1987-1988-1989; Algemeen Bestuur Ontwikkelingssamen Wering, Brussels, 1990.)

38. The countries are Angola, Cape Verde, Comoros, Chad, Equatorial Guinea, Ethiopia, Guinea Bissau, Guinea, Malawi, Mozambique, Nigeria, Sierra Leone, Sudan, and Uganda. The countries with especially heavy UNDP assistance are Malawi and Ethiopia.

39. Although this is true, in most cases technical cooperation expenditures are not accounted for in government budgets.

40. République du Tchad, Ministère du Plan et de la Coopération, "Etude diagnostique de la coopération technique au Tchad," février 1990. Study carried out as part of NaTCAP exercise.

41. "Thanks to this realistic and progressive approach, we are convinced that 10 or 20 years from now we will no longer require expatriate technicians."

3

IMPROVING THE DELIVERY OF TECHNICAL COOPERATION

Many factors lie behind the poor track record of technical cooperation for capacity building. One set of factors consists of systemic or macro-level problems that undermine the impact of technical cooperation: weak national management of this form of aid, the lack of an effective "market" for technical cooperation, and civil service disarray in recipient countries. These problems are mentioned only briefly here; they are taken up in detail in later chapters.

A second set of factors consists of persistent errors or shortcomings in the identification, design, and implementation of technical cooperation projects. These are addressed in this chapter. The first part of the chapter summarizes the main lines of what has become a consensus diagnosis; the second considers proposals for reform.

FAILURES IN DELIVERY SYSTEMS

There are two main types of failure: (1) inadequacies in dealing with the nuts-and-bolts problems of devising, delivering, and monitoring sound technical cooperation projects—that is, projects that are responsive to priority local needs, tailored to local institutional capacities, implemented with dispatch and by the right people, and monitored to ensure their responsiveness to changing circumstances; and (2) excessive reliance on one instrument or model for technical cooperation delivery: the expatriate resident adviser with local counterpart.

Defects in Identification and Design

Deficiencies of design have been catalogued in Chapter One, along with other criticisms of technical cooperation. In report after report, and particularly in the recent NaTCAP assessments, the refrain is the same. Technical cooperation projects are often not aimed at the highest-priority needs. They do too much substitution for local staff and not enough capacity building. They reflect local preferences imperfectly—preference, for example, for local as opposed to foreign experts or for more training and fewer experts. The projects are often too ambitious and use inappropriate technology or approaches.

Lack of Local Involvement

One underlying reason for many of these defects is the lack of local involvement in identification and design of projects. Technical assistance is too much and too often a donor operation. The conception of most projects originates within donor agencies, whose identification missions flesh out initial ideas. The design of projects is the work of donor agency staff or consultants and consulting firms hired by the donor, as is their evaluation.

Profound ramifications result. Lack of local participation in identification and design leads to low degrees of recipient government ownership and hence low commitment to many projects. Also, technical cooperation projects are selected largely ad hoc, with each donor-supplier acting independently. Central control and coordination are limited, often nonexistent. There is little or no systematic attention to ordered national needs, or to the linking of technical cooperation to priority objectives of the recipient government. Some needs are lavishly served, others neglected.

This basic and pervasive factor—the donor-driven character of so much technical cooperation—is a root cause of the ineffectiveness of this form of aid. Its negative consequences are considered in detail in Chapter Four.

Overly Complex Projects

Many technical cooperation projects are overly complex. This derives partly from donor failure to draw on local experience and insights, and partly from the many objectives that the donor-designers seek to satisfy with technical assistance. Technical cooperation sometimes seems to serve donor strategies, aid objectives, and internal process requirements as much as it serves the development needs of the recipient countries.

- Donors want expatriate personnel to ensure expeditious implementation of "their" capital projects within which technical assistance is incorporated. They believe also that having these personnel raises the chances of success in capacity-building components and projects. This position derives from donor uncertainties about the availability of adequately trained, experienced, and motivated local staff.

- Donors see expatriate technical assistance personnel as helpful or essential in the performance of sensitive aspects of project management—the control of project budgets, for example, or the monitoring of conditionalities in policy loans. Also, since technical assistance personnel are usually more knowledgeable than local personnel about home office reporting requirements, their use lightens the administrative burden on resident aid missions. It is not surprising, then, that the Nordic study evaluating technical cooperation in Kenya, Zambia, and Tanzania classified over 17 percent of the technical assistance personnel in those countries as essentially controllers, while 65 percent were implementers, and only 11 percent concentrated on training and 7 percent on institution building.[1]

- The demand for expatriate technical assistance is inflated by some idiosyncracies of project design. Often, project designers set about their task with little or no reference to local institutional capacities, and with technical efficiency the main guideline in shaping the project. This approach leads to the old and deep-rooted propensity to design complex or high-tech projects and then conclude that expatriate assistance is essential for project implementation.
- Related to this is the propensity among some project designers to assume away difficult institutional problems by asserting that they will be dealt with by additional technical cooperation. This is common in many important reform areas such as strengthening of public enterprise management or public expenditure programming. Appraisal reports or project papers outline the formidable deficiencies of current practice, then propose technical cooperation as the principal remedy. In this and other ways, institution-building targets are consistently overambitious and unrealistic.

Hurried Design of Components and Projects

In most instances, little time is given to preparation of technical cooperation components and projects. In investment projects that include technical cooperation, the capital component is top dog; it very much wags the technical cooperation tail. The design of hardware—roads, power plants, dams, and factories—receives much more attention than the technical cooperation that may be called on to help in construction or in initial operation of new facilities.

One part of technical cooperation—the training component—tends to receive especially short shrift. Training needs are almost always given less careful attention in project preparation than the other components. This is especially true of training components in investment projects, but it is also true of free-standing projects focused on identification. World Bank evaluators have dubbed this phenomenon "training as afterthought." It is common in technical cooperation projects of all donors. Many evaluations have pointed out that project designs are usually very sketchy on the training function and that terms of reference treat it vaguely.²

Poor Drafting of Terms of Reference

Scopes of work are often drawn up with inadequate care.
- Job descriptions are too broadly or too loosely defined. They not only provide little guidance to incoming personnel but also rarely include measures of performance or output that would allow effective monitoring and evaluation.
- Most important, the capacity-building, institutional-development objectives of the projects are often muted and rarely given highest priority. Terms of reference are usually not explicit enough on the ranking of priorities to truly orient or constrain behavior of expatriate personnel. The authors have seen no terms of reference that state unambiguously, for example, that the volume or quality of other work should be sacrificed if necessary in pursuit of capacity-building goals.

In part, this failure to give strong explicit priority to institutional-development objectives reflects lack of agreement among all the actors on the objectives and uses of technical assistance personnel. Some user agencies give high priority to gap-filling or personnel substitution objectives. In the past, at least, these agencies were unwilling to see these objectives explicitly disavowed in terms of reference.

A frequent consequence is that long-term resident experts who are supposed to act as advisers are assigned staff or line responsibilities in the user institution whose leadership is often more interested in short-term outputs and accomplishments than in long-term capacity building.

Disadvantages of the Project Mode

The encapsulation of technical assistance within projects has certain built-in inconveniences that have limited its long-term impact.
- The reliance on ad hoc and autonomous technical cooperation projects makes difficult the introduction of more integrated and

coherent approaches based on the "programme" concept—the idea that problems should be defined and attacked in sectoral or even wider terms. Problems of village water supply, for example, may not be dealt with best by introduction of an uncoordinated, independently defined set of technical cooperation projects—one here to improve well-digging capacity, another there to strengthen well maintenance in a different region.
- Project-related technical cooperation tends to move on a time path that is too short (two to four years) given the intractability of many of the institutional bottlenecks being addressed.

Weaknesses in Implementation

The range of implementation problems that reduce the effectiveness of technical cooperation and raise its cost is wide. It is useful to distinguish systemic problems from those specifically related to implementation.

Systemic Problems

Some of the most important sources of the ineffectiveness of technical cooperation derive not from particular projects or specific project delivery arrangements, but from systemic factors—notably, weaknesses in the general administrative and work environment. Local staff often lack monetary incentives to work hard and commit themselves to the job. Budgetary and related constraints make for unsatisfactory support arrangements (unavailability of government counterpart financing for supplies, travel, office equipment, and so on). Coordination and general management of the technical cooperation process are weak. These factors are considered in Chapter Four.

Other problems in the same category include delays in procurement of project equipment and supplies because of burdensome procurement regulations and lengthy customs-clearing arrangements, absorption of expatriate adviser energies in such housekeeping details as finding suitable housing and ensuring maintenance and repairs, and lack of adequate supervision and quality control by the financing agency.

Implementation problems of a somewhat different order arise from negative spillover effects of the presence of technical assistance personnel. The search for counterparts and the incentive payments donors dispense to local project staff create a dual salary structure within the public sector and generate resentments among the nonfavored, with probable withdrawal of effort as a result (see Chapter Six).

The expatriate presence generates other negative external effects. Many local staff resent the high salaries and opulent living of expatriate technical assistance personnel. Many doubt that the experts' superior competence merits the salaries and other benefits they receive. Local staff feel excluded from the expatriate networking that is so common in many administrative systems and that provides much easier access to information than is possible for local staff. In all these respects, the expatriate expert presence has unsettling, intangible effects on organizational performance.

Implementation Problems Specific to Technical Cooperation Projects

Here we are concerned with implementation problems that are more specific and operational: delivery system deficiencies that are internal to the projects themselves. Some of these are universal. All donors, for example, have difficulty recruiting suitable expatriate technical assistance personnel in a timely manner. The requirements in technical and language skills, experience, education, personality, family situation, and availability are extremely demanding. It is not surprising that recruitment delays and mismatch of personnel with local requirements are widespread.

Inadequate monitoring and supervision is another source of weak performance. Although this is a problem in all development assistance projects, it is especially prevalent in technical cooperation projects. These projects are often less amenable to midstream assessment than are capital projects. To put on track a floundering technical cooperation project usually calls for a change in the cast of characters, which is difficult to bring about. And because the presence of expatriate personnel raises so many issues of working relationships and conditions, supervision tends to focus inordinately on procedural matters.

We review these and other implementation problems in the context of the general delivery model that has dominated the technical cooperation process in Africa—the resident expert-local counterpart model.

The Resident Expert-Local Counterpart Model

The technical cooperation model that has gained widest currency in the developing world is based on the use of long-term resident expatriates (experts, advisers, *coopérants*) who are to train and transmit technology mainly via a one-on-one relationship with a local counterpart. Resident expatriate personnel have been and continue to be the mainstay of technical cooperation in Africa.[3] In 10 countries for which data are in hand from NaTCAP studies, personnel costs accounted for between 53 and 86 percent of total programme expenditures.

Central to the issue of the expert as a mode of technical cooperation is the concept of the expert-counterpart system, which underlies the policies and practices of most multilateral and bilateral donors. Four main elements make up this system: (1) a national counterpart must be assigned to work with each expert; (2) he should work full time, as a kind of understudy, with the assumption that he can take over the job when the expert leaves and the project is finished; (3) the expert should act as an adviser and not fill a line post in the administration; and (4) the task of the expert is to train the counterpart so that he can function independently.

The expert-counterpart model is now widely condemned, although in practice it continues to be the principal vehicle for technical cooperation in Africa. Much of the recent criticism of the effectiveness of technical cooperation focuses on one side of the model—the activities of experts and the continued reliance on expatriate personnel. In donor circles as well as among recipient country officials, there is a widespread rejection of the resident expert modality of technical cooperation; UNDP, OECD, and the World Bank as well as many bilateral donors have called for reduced reliance on this instrument.[4] And statements of policy on technical cooperation issued by many African governments call for a shift away from resident expatriates to short-term consultants and national experts.[5]

The reasons for the failure of the resident expert-counterpart model are well known and well documented, particularly in recent studies and policy statements resulting from NaTCAP exercises. The expert concentrates on getting the work done rather than on training, is often good at his job but bad as a trainer, upstages the counterpart in influence, and sometimes blocks the counterpart's career progress by staying too long. Counterparts are too few and often not right for the job, are selected too late in the life of the project, are too lightly trained, or quit for better jobs.[6]

Basic Preconditions Unmet

It is clear that the expert-counterpart model of technical cooperation rarely works in practice as it is supposed to work in theory. The model is weakest precisely in achieving the objectives of building national capacity through training of nationals and through transfer of knowledge and technology. The basic conditions for reasonably successful operation of the model are usually not met. These conditions were mentioned above: the presence of suitable national counterparts and acceptance by both partners of the understudy relationship.

Reasons why the assignment of national counterparts is a nearly universal problem in Sub-Saharan Africa have already been noted. Counterparts are not selected at all or are selected too late in the life of the project, are inadequately qualified or very junior, are not assigned full time to the project, often have little material incentive to work at the assigned job, and in any case often quit for better jobs elsewhere.

In a significant number of cases, the arithmetic is constraining: there are simply too few nationals in post and available for counterpart assignments. The assumption underlying the model—that there are nationals in place who simply lack on-the-job experience and training—does not hold. In Malawi, for example, a recent survey of the agriculture ministry revealed vacancy rates (unfilled established posts) of 25 to 50 percent in the higher professional and technical grades.[7] There are even a few cases in Sub-Saharan Africa where the number of expatriate personnel exceeds the number of national civil servants at professional levels: in Guinea-Bissau in 1987 there were 489 full-time resident expatriate technical assistance personnel, about 100 more than the 398 national civil servants with university degrees.[8] Even when more trained nationals are available, managers who are short of staff are reluctant to assign competent individuals as counterparts because, if faithfully done, this results in two people doing one person's job.

The model assumes a public sector work environment conducive to one-on-one transfers of skills and technology. But national counterparts often have little incentive to perform, even to come regularly to work, because of low salaries and bad working conditions. They have even fewer incentives to be interested in job-specific training, because their future income is almost always maximized by changing posts. The general work environment, in addition, is often uncongenial to the business of serious training: management is lax and unstable, delegation of authority is hesitant, discipline is weak.

The understudy role of the counterpart is rarely applied because the counterpart often leaves when the project finishes. Indeed, often it is the other way around. The project collapses when external financing is ended. Nor is the advisory role of the expert respected. He or she lapses easily, usually enthusiastically, into an operational role. The same is true, even more decisively, for the training role. This is so even in those (relatively rare) cases when terms of reference are clear and strong in favor of institutional development and training. Powerful forces pull expatriate personnel in the direction of producing other, non-capacity-building outputs.

The internal values of the expert himself push him to produce hard outputs. Expatriate personnel are selected largely on the basis of technical competence, after all, and successful performance on the job.

They are not accustomed to background roles, to letting others do the work, or to gently counselling peers or subordinates.

- The ground is often infertile for capacity-building efforts—lack of appropriate counterpart staff, instability in leadership, and lack of budgetary support, for example. It strikes many as a low-yield activity and of dubious priority when there is so much urgent work that needs to be done.
- Hard outputs are usually what the bosses of the expatriates really want, and their performance and worth will be judged mainly by their ability to produce those outputs quickly and well.

Recruitment Problems and Divided Loyalties

Recruitment difficulties and delays plague most technical cooperation projects, in particular those relying on long-term resident personnel. It is always hard to find the right person—someone willing to travel or relocate who has the right mix of technical, educational, language, training, and interpersonal skills, and the needed experience, formal training, cultural knowledge, temperament, and family circumstances. It is harder still to find the specialist at the right time—which, by common practice in the technical assistance business, is usually right away. The task is complicated still further the higher the priority given to capacity-building and training objectives. The work habits and instincts of people who are good technicians or doers are not often the same as those required of good trainers. And these qualities are much more difficult for recruiters to assess.

Recruitment difficulties are exacerbated by a common tendency to impose inflated formal requirements for the job—many years of work experience, knowledge of the country or region, advanced degrees, high language competence, and so on. Such profiles are discouragingly rare. It is not surprising that technical cooperation projects are often slow in start-up and understaffed for many months, and that there is widespread recipient country dissatisfaction with the quality of experts finally recruited.

Some expatriate personnel may face conflicts over split loyalty and accountability. Their allegiance is inevitably divided. They work for and report to the leadership of the host country institution within which they are employed. But they also are involved with the donor institution. And where there is an intermediate supplier, say a nongovernmental organization or a consulting firm, there can be three bosses. In principle, and in formal terms, the experts' prime loyalty is to the local agency for which they work. For most technical assistance personnel this presents few difficulties. But their careers depend much more on how they are evaluated by the donor agency, or by the consulting firm or nongovernmental organization that hired them, than by the user agency's assessment. So the potential for conflict does exist. And in many cases these ambiguities in lines of authority complicate the task for local managers seeking to impose coordination and control.

An Inappropriate Model for Training

Finally, and most fundamental, the expert-counterpart system is an artificial and untested training vehicle. It can be found nowhere else but in the world of technical cooperation. No public or private organization has developed institutionally or trained its staff through such an instrument. Even aid agencies do not use this approach in training their own staffs or in developing their institutional capacity.

As on-the-job training works in most organizations, the trainee is given clear operational tasks and responsibilities under the supervision of a more experienced professional who can check on and guide performance. This is the way most company training programmes for new management recruits work, and it is how international institutions and aid agencies do their training—for example, the Junior Professional Officers in the United Nations, the Young Professional Programs at the World Bank and the International Monetary Fund, and the International Development Intern Program of the United States Agency for International Development.

This is not how the expert-counterpart system works. Under this arrangement, the principle of learning by doing is violated because the counterparts cannot work independently and learn from their own successes and failures. They are inevitably overshadowed by the experts and rarely given full responsibility for project activities.

Issues of hierarchy and personal relationships are also treated differently in the expert-counterpart system than in other on-the-job training systems. In most trainer-trainee situations, lines of hierarchy are relatively clear and tensions in personal relationships are not usually much different from those found in most work relationships. In the expert-counterpart arrangement, however, an unspoken assumption prevails that the resident expert and the counterpart are professional equals, distinguished mainly by degrees of experience. The expert is supposed to "advise" his partner-collaborator, helping him do his work in a collegial way.

The reality of course is otherwise, and numerous contradictions arise as a result. Although the experts are supposed to only advise, they are older, more experienced, vastly better paid and motivated, better equipped with transport and computers and paper for copying machines, and often plugged into an expatriate network within which information flow is much more abundant than anywhere else in government. Moreover, they often hold the purse strings and the key to all project equipment and materials while these resources are very much is short supply elsewhere in the administration.

Because the experts can usually produce better work and turn it out faster than their counterparts, local managers tend to rely on them directly. Thus although in theory an adviser, and not formally operational or part of the administration, the expert often has access to senior officials in the ministry and outside, and possibilities for influence that are beyond the reach of the national counterpart. The counterpart can rarely manage the project activities because only the expert has the means to do so.

A final flaw in the expert-counterpart model as training vehicle relates to cultural factors. Established hierarchy and effective on-the-job training are possible in systems where the trainer understands both technique and corporate or organizational culture better than the trainee. On-the-job training transfers skills in a real world rather than a classroom context. An inexperienced staff member can learn from a senior professional because the latter has professional competence and superior knowledge about how the world works. In the expert-counterpart situation in developing countries, however, the organizational and cultural context at the work place is not the expert's own. The national counterpart understands the local context better than the resident expert. The expert's credibility and effectiveness as a trainer and institutional developer are thus diminished.

PROPOSALS FOR REFORM

Recall at the outset that in this discussion we take as given the macro setting, the basic structural characteristics of the technical cooperation "system": its management by donors for the most part, its peculiar "market," and the fact that the public sectors within which most technical assistance personnel are inserted suffer from lack of incentives and generally poor working environments. We assume that these are unchanged or that they are being separately addressed. Two sets of reforms of technical cooperation delivery can then be considered: improving the way the nuts and bolts are put together to make the existing arrangements more efficient and effective; and making changes of a more basic kind, notably by substituting other forms of delivery for the expert-counterpart system.

Delivering the Existing Package More Effectively

The key words here are: "Do it better." Address the deficiencies acknowledged by every report and admitted by virtually every practitioner.
- On design: give more time to technical cooperation project preparation, seek local participation; take account of local institutional capacities and real needs, avoid automatic resort to top-of-the-line technology, don't include too many complex components, give more care to training components, avoid too loose terms of reference, specify measurable outputs in these terms of reference, emphasize capacity-building objectives more

than operational support, and specify lines of responsibility for technical assistance personnel.
- On implementation: improve recruitment procedures to reduce delays; pay more attention in recruitment to cultural sensitivity, interpersonal skills, and training capacity; work harder to ensure presence of counterparts; and improve supervision and quality control by donors.

These proposals have one particularly striking characteristic: they are heavy on exhortation, light on how these desirable changes are to be brought about. Their grammar is long on "shoulds" and "musts." A few direct quotations from the 1991 DAC report on principles give the flavor of such proposals.[9]

> Participation of beneficiaries in the identification, planning, implementation and evaluation of all kinds of development assistance projects, in particular Technical Cooperation, is essential to ensure that they take on the full control and responsibility of activities and pursue them after the donor's departure, a key element of sustainability.
>
> Donors and international agencies should exercise restraint in offering their own isolated TC proposals.
>
> Greater attention needs to be given to cost effectiveness.
>
> The performance of TC experts can be improved through more careful recruitment, better briefing on the socio-cultural, economic, political and institutional environment of the recipient country, stronger technical backstopping and, above all, clearer terms of reference. Donors should select experts not only for their professional competence but also for their ability to exchange and transfer experience.
>
> Re-examination of the 'counterpart/expert' arrangement should aim at establishing a true partnership, where partners are considered to be equal and where each has clearly defined functions and responsibilities. . . . Training of counterparts should be given necessary emphasis so that they may take over full responsibility within the shortest possible time span.

Two questions arise in assessing these mostly sensible appeals for better performance by donors. First, at least one recommendation may not be correct. This is the insistence on tighter terms of reference and the use of measurable output indicators to judge technical assistance personnel performance.

One source of doubt about this blueprint approach is that it runs counter to everyday experience. Consultants rarely find that terms of reference fit the problems they discover when they arrive in country, and even more rarely are they of much help in getting the job done. A perceptive observer puts it well.[10]

> To write detailed TORs requires pretty comprehensive understanding not only of the problem but how it should be solved. Do we really know enough to map out the TA work in such detail? Is there not a danger that requiring too specific TA will force the project into the straitjacket of replicating the methods used in another country? This may be fine for designing a road, but not for strengthening a planning system or improving civil service management.
>
> Looking back on my own TA experience (15 years in a range of countries) I cannot think of a single case when the TOR was of any use in guiding me how to tackle the job. You read them on the plane out, and out of convention refer to them in the introduction section of your report. For the rest of the time they remain in the drawer. Why not recognize the reality that TA tasks cannot be programmed in detail, that advisers must respond to opportunities as they find them . . . and that with ID TA however careful the preparation the best way to do the job can't be specified in advance.

The second and more basic weakness of these exhortations to do better is that they take insufficient account of the profound causes of the deficiencies in question. The problems of hasty project design, excessive complexity, low training yields from the expert-counterpart model, or poor supervision are not due to wickedness, stupidity, or lack of knowledge. After all, the recommended changes are well known and have been openly embraced by almost everyone. They have been put forward repeatedly by the authors of the numerous reports on technical cooperation that have appeared during the past 10 to 15 years. Yet only minimal changes are observable.

The reason is that these deficiencies reflect—in addition to basic structural flaws in the technical cooperation system (a donor-dominated environment, no constraining market forces, and inhospitable local work environments)—deep-rooted habits of organizational behavior on the donor side. These organizational behaviors are not readily changeable. They are quasi-genetic, if analytic license allows use of a biological analogy. When reformers call for better performance on these matters of project design and implementation, they are really asking organizational leopards to change their spots.

A revealing example is the excellent recent World Bank study on technical assistance and institutional development.[11] In a chapter titled "Why Is Our Record So Poor," the author describes some of the World Bank's built-in handicaps as a provider of technical cooperation for institutional development. The organizational characteristics she mentions are:

- A blueprint approach is taken to project design and implementation; this is unsuitable for capacity-building technical assistance projects, which are better advanced by an adaptive or process approach to implementation;
- There is a lack of institutional memory and cultural awareness, caused mainly by high staff turnover on projects;
- Staff and management give priority commitment to project efficiency, not to identified goals; this is exhibited, for example, in continuing creation of project implementation units long after their negative institutional impacts have been recognized, and in payment of salary supplements despite disruptions they cause in local salary structures.
- Priority attention is commanded by capital lending and getting projects to the Board, leading to neglect of technical assistance projects that are complicated and take a long time to prepare.
- Supervision is something of an orphan for all projects. Its career payoffs are small. The staff is heavy with economists

and other technocrats who are better at analysis than management and do not like to go on supervision missions in any case.
- Government commitment is given slight attention because of pressures to lend, because commitment is hard to analyze and measure, and because technicians are uncomfortable with touchy-feely subjects like assessing motives.

Any organization can do better, in technical cooperation as in other activities. But with basic characteristics like those described, it is easy to see why actual improvements lag so far behind recognition of deficiencies. It is also legitimate to wonder whether the organizational culture is such that a faster rate of improvement can be counted on in future. What is true of the World Bank is largely true of other aid agencies. This means that patching up the standard technical cooperation package is not enough. More fundamental changes are needed. Other paths have to be discovered, other instruments unsheathed.

Changing the Mix of Delivery Modes

It is certainly possible to patch up the expert-counterpart model, to remedy some of its obvious deficiencies—for example, by better specification of training roles in terms of reference and by strongly and explicitly emphasizing the training and capacity-building function. But the resident expert-counterpart model has revealed flaws that are so general and so fundamental that they put into doubt its salvageability. No amount of exhortation, no tinkering with terms of reference seems capable of transforming that ill-conceived training model into an effective vehicle for capacity building.

The central reform proposition therefore is to reduce drastically, perhaps abandon altogether, reliance on the expert-counterpart model. It should be replaced by renewed or fresh emphasis on four substitute arrangements: an open use of gap-filler types of technical assistance personnel, a wider adoption of coaching models that rely on short-term personnel and repeated visits, greater reliance on local consulting capacities, and wider resort to twinning.

Unashamed Gap-Filling

Most OECD donors have abandoned or are abandoning gap-filling, or provision of expatriate personnel in operational roles. It has fallen out of fashion in current aid philosophies. But other donors have used it in the recent past and some continue to do so: the Chinese, Russians, Egyptians, Cubans, and Algerians, for example.

A real need exists for this type of assistance; in many African countries, educational outputs in professional and technical fields still fall short of needs, even modestly defined. Thus out of a total of 300 medical doctors in Burundi, 100 are foreigners working in operational posts. Many technical assistance personnel also fill a substitution role but pretend to be advisers. It would be better to be honest and recognize these needs for operational staff.

The necessary conditions for making these operational staff effective are to integrate them into the national administration. The old British topping-up system for expatriates filling line positions and the comparable United Nations' OPAS system seem sound for two main reasons. First, the positions that are filled this way are established posts that are vacant. They are thus provided for in the civil service roster and, in principle, in the budget. In theory, after the technical assistance person leaves such a post, there would be no administrative or financial constraint to continuing the position and the activities of its holder. Second, the expatriate is integrated better into a national hierarchical structure. He or she is clearly part of a team and works with nationals as a peer, a subordinate, and a supervisor rather than holding that ambiguous position of adviser cum counterpart.

It does not seem accidental that countries operating with this system exhibit better management of their technical assistance. Papua New Guinea, for example, has some 1,500 contract officers in government, occupying line posts. This technical assistance seems well managed compared with that in most African countries.

More relevant, this is the system that has long been used in Botswana, the country with probably the best-managed technical cooperation inflow in all of Sub-Saharan Africa. An economist who worked as a resident expert in Botswana described the arrangements this way.[12]

> When I worked in Botswana in the 1970s the bulk of TA was individuals on local contracts with their salaries topped up by ODA. We occupied line positions, had our posts recorded in the Establishment Register, were subject to civil service discipline, and signed the Official Secrets Act. Local accountability of TA was taken even to the point of sending expatriates to jail. (One TA person was given a stiff jail sentence for fiddling with his travel claims.) At budget time, the issue was not whether technical assistance should be given to a particular ministry, but whether a new established position was justified. The merits and demerits were argued in that context. How the post was filled was a matter of implementation. If no suitable local was available, the Department of Personnel would approach one of the bilateral donors, or UNDP.
>
> We had a firm policy to avoid technical assistance 'projects' (of the conventional type). TA people were to fill line positions, continuing in them until a local was ready. There were very few advisers—only a select number of very high level policy types, mostly in the Ministry of Finance and Development Planning. We held that scattering TA advisers across the government organizational structure would create conflicting lines of responsibility, and undermine ID. We were skeptical about counterparting, believing that the best way to promote transfer of knowledge was by having expatriates and locals work alongside each other in line positions. . . .
>
> Once or twice we did agree to TA projects, and regretted it. I asked (one donor's) help in providing a couple of agricultural economists for the MOA planning unit. Instead of accepting the government's standard job description for the post, they sent a team to write up a long project document, greatly overblowing it. An advisory team was duly established in the MOA, but it never really worked; (they) were never properly integrated into the cadre or became part of the planning officer network that was

part of the glue that holds the Botswana government's development performance together.

Gap-filling of course is not free of problems. It should not be used excessively, and all parties involved should understand clearly that this kind of personnel, like others, has to be replaced on a phased basis. Governments have to prepare the transition by formal and informal training programs. But the chief problem of the gap-filling approach is not that expatriate personnel will stay too long. It is rather the sustainability of the approach in the face of the need to reduce public sector employment. Filling established posts would normally go a long way to ensuring that expatriate personnel are truly needed. But many African governments cannot afford to pay the salaries of existing civil service establishments; many posts, vacant or filled, will have to be eliminated.

Greater Use of Short-Term Advisers and Coaching Models

This is now standard doctrine. The DAC's 1991 principles report states it clearly.[13]

> Although experts on long-term assignments in advisory rather than operational roles may be needed and thus still requested by recipients, the use of expatriate professionals solely for project construction and operations should be exceptional. . . . More emphasis should be given to short-term experts including more frequent follow-up visits.

The pure coaching model involves short-term visits by expatriate technical assistance personnel working with national staff on specific assignments. After an initial mission, which might be longer than usual (for example, two months rather than the standard two to three weeks), the expatriate expert helps define the problem and begins to attack it, with local staff. He then leaves, and responsible national staff carry the activity forward. The coach remains in close electronic contact with the project, perhaps through a paid national administrative person, and returns every few months to discuss progress and problems with the responsible national officials.

A modified version of this approach has a much more traditional look. This is to have one long-term resident project manager, usually an expatriate, who provides advice and—under the direction of local officials—defines needs for short-term consultants and organizes their missions.

The problems of the pure coaching model are easy to see. Unless the local officials engaged in the activity are, first, truly committed to it and, second, able to resist diversions to other work, not much will happen when the coach goes home. Also, since the surrounding environment is always changing (cabinets are reshuffled, officials leave their jobs, drought hits, or civil disturbances erupt), the work programme established during the last visit has to be recast and other adaptations made.

The move to greater reliance on short-term consultants is hindered also by its impact on donor organizations. Most donors are set up to do recruitment of long-term advisers in the framework of technical cooperation projects. Use of short-term people creates new demands and new administrative burdens for these suppliers. For example, in the UNDP system, the Office for Project Services now prepares Requests for Proposals and works through the process of competitive bidding. None of this is needed for intermittent consulting types of hiring.

Bilateral donors that work extensively through contracting out to consulting firms would be less troubled by a deeper shift to the coaching modes. Their procurement operations are already set up to do recruiting and (less clearly) supervision of short-term intermittent consultants.[14] But significantly greater reliance on this type of personnel might require new management structures for recruiting and, in particular, greatly strengthened backstopping capacity in many aid agencies. This approach might also require stronger resident missions.

The important points are two. First, intermittents are much more demanding of donor administrative support and hence more expensive. Second, the more intensive use of intermittents would require changes in some donor structures and moves into unfamiliar administrative territory. For these and other reasons, the movement away from the resident expert-counterpart model could be slow.

Greater Use of Local Consultants

This has been a standard element in virtually all technical cooperation reform proposals for a decade or more. Donors have accepted it in principle, and in fact many individual African consultants and the few African consulting firms that have developed in recent years are increasingly solicited by donors.

The broader use of local consultants should be a central part of any capacity-building effort. What it takes to do this is well known. The DAC report contains a good summary statement.[15]

> The use of developing country consultants by donor consulting firms, including joint ventures, should be encouraged and efforts made to enable developing country consulting firms to compete with donor country firms. Donors could: i) support training programs for recipient country consultants; ii) encourage twinning arrangements between donor and recipient country consulting firms; iii) help recipient governments assemble information on local expertise; and iv) assist newly established recipient country firms to become known to the international market.

Despite such sound counsel and nearly universal recognition of the need to nurture local consulting capacity, this capacity remains embryonic in many, probably most African countries. Its slender presence is the main reason for the limited contributions of local consultants in most of the region.

This is not everywhere the case. In some countries, local consulting capacity is substantial and growing. Ghana, Madagascar, Nigeria, Kenya, and Zaire stand out. Ghana, for example, has for a long time had a significant stock of highly educated and experienced technicians and professionals, and a low-wage public sector. A private consultant sector of some size has emerged. A 1987 study identified 390 experts and 60 consulting firms. Most of the firms were small (one to two people) and young (formed since 1985). But 76 professional economists were inventoried, about a quarter of them with Ph.D. degree training. And many experienced auditors, accountants, and engineers are available. This is certainly raw material for a private consulting industry with real potential.

Several obstacles to the expansion of local consulting capacity and its more intensive use have been identified. Probably most important are inertia, administrative regulations, and political pressures that induce donor representatives to contract with home country consulting firms. But other factors are at work. Some donor spokesmen worry that use of consultants will accelerate the flight of competence from the public sector and allow postponement of hard decisions about civil service reform.[16] The limited experience of many local firms gives other donors pause. And, in some countries, local political pressures in favor of specific local firms can be counterproductive.[17]

Institutional Twinning

Twinning is another hallowed idea. More than 15 years ago, UNDP reported on experiments in institutional twinning it had undertaken. The purpose of twinning, stated the UNDP document, is "to enable the assisted institution to gain access on a systematic and continuing basis to know-how accumulated in the other institution and—ideally—to develop a more permanent partnership from which both institutions will profit for an extended period of time." The report pointed out that twinning between universities was a common phenomenon in bilateral assistance for many years.[18]

Efforts to pair organizations with similar operational functions have not been limited to universities. Building on colonial experience, international units of British institutions, including the railways, postal service, water authorities, and electrical utilities, have contracted to transfer technology to counterpart institutions in developing countries.[19] In 1975, the UNDP Governing Council specifically cited the International Water Supply Association and the International

Association of Ports and Harbors for their role in developing "new dimensions in technical cooperation, "by facilitating twinning among related agencies.[20] Consulting firms in industrial countries are increasingly forging twinning relationships with firms in developing countries.[21]

Thus, twinning as a delivery mechanism for technical assistance has long been used to meet the evolving priorities of the donor community. But the number of sectors that have experimented with twinning is far greater than the number of times twinning is chosen as a form of technical assistance.[22] When compared with the use of other delivery mechanisms, primarily long-term expatriates and training, twinning accounts for only a small percentage of technical assistance. In a World Bank review of 135 projects approved in fiscal years 1987 and 1988, for example, only eight included definite plans for twinning, while twinning arrangements were being pursued in an additional four projects.[23] By the most generous interpretation, the World Bank's use of twinning as a form of technical assistance is a mere 8 percent. Additional reviews reveal few concrete examples of twinning.

An inventory of sources of twinning prepared in 1984 identified fewer than 10 specific institutions.[24] In 1983, the Swedish International Development Authority (SIDA) had worked with only one institution, Statistics Sweden, on just seven twinning arrangements.[25]

Despite the paucity of experience, twinning has a strong reputation as an effective form of capacity building. In some circumstances its promise is certainly great. When the twinning partners, for example, draw mutual benefits beyond the contractual, the basis for effective and sustainable cooperation exists. These are instances where the supplier perceives its gains to be greater than straight cost reimbursement. Examples may clarify this point. When an environmental organization in an industrial country (for example, the U.S.-based Nature Conservancy) twins with environmental nongovernmental organizations in developing countries, it draws two special benefits. First, any conservation successes achieved by the indigenous nongovernmental organization contribute to its objectives. And, second, the staff in the industrial country organization benefits from increased knowledge of environmental conditions elsewhere in the world.

This mutuality of interests does not play the same role in profit-making organizations. Here the services have to be contracted for on market principles. The advantages of twinning thus have to be based on cost-effectiveness considerations.

There are good reasons to anticipate greater effectiveness from twinning than from more traditional modes of technical cooperation delivery. First, the shared technological environment and overlapping business preoccupations create a large and immediate zone of common interest and common understanding. In the case of power company twinning, a British supplier noted the existence of "a high degree of compatibility between the client's technical requirements and the developed utility's own experience." The result is a far greater acceptance by the recipient country of the credentials of the overseas partner, and hence greater credibility of advice proffered.

Because shared expertise is a characteristic of twinning, however, it may not be appropriate in instances when the recipient needs a wide range of expertise. In two projects the World Bank sponsored in Ghana, for example, the Ghanian government was reluctant to lock itself into twinning relationships because it was afraid to confine the assistance to the experience of a particular company.

A second advantage of twinning relationships is that interaction occurs among several individuals, rather than among a limited number of experts and counterparts. Because twinning can involve study tours in the supplier's institution and numerous types of short-term assistance from the supplier, more participants interact at more levels, leading to a more pervasive institutional impact. Advisers and counterparts in the traditional mode turn over frequently, diluting their capacity-building pay-offs. By spreading impacts more extensively, twinning reduces the risks of dissipation of capacity when experts and counterparts disappear. As one World Bank task manager said of twinning, "The seeds go a lot further."

A third advantage is the adaptability of twinning relationships. Reviews of the effectiveness of technical assistance indicate that unpredictable factors such as personnel changes and shocks of various kinds can derail the most carefully outlined project objectives. In twinning relationships it is relatively easy for the parties to reassess their circumstances and renegotiate goals.[26]

Finally, twinning can result in long-term relationships, implying greater effectiveness. Many twinning relationships initially establish a time frame longer than the normal technical cooperation project. Twinning partners may be able to extend the relationship more easily than the rebidding of normal projects.

Counterbalancing disadvantages exist. Suppliers may not be plentiful. It is not clear that incentives exist for industrial country profit-making organizations to engage in new and unfamiliar activities far from major markets for modest returns. Aid agencies seeking to encourage participation may have to invest substantial resources to guide and encourage the overseas supplier until it can operate effectively overseas.[27]

Twinning is not good for everything. When an organization's activities are diverse, it may not be appropriate. It may also be difficult for parties unfamiliar with these arrangements to define mutual objectives clearly and to agree on them. A World Bank-sponsored project in Guatemala is plagued by disagreement over the diagnosis of the needed technical products and training. The project involves a twinning arrangement between the Guatemala City water company and a consortium that includes a Brazilian water utility and three engineering firms (Brazilian, Israeli, and American). In addition, the budget allocated for training and exchanges is inadequate, and the selection of training candidates is politically driven.[28]

The greatest deterrent to establishing an effective twinning relationship is the time and effort required to find two institutions that will make the right match. The infrequent resort to twinning, despite its universal appeal, suggests that this is not an insignificant problem. Also, the costs of guessing wrong on a partner are much higher in twinning than in other delivery arrangements. An unsuitable resident adviser can be put to work in a quiet corner. Inappropriate twinning—like a bad marriage—is much more difficult to ignore.

With regard to costs, it is not apparent that twinning is less expensive than comparable forms of delivery—long-term resident teams, for example. One evaluation, by the Swedish aid agency, compared estimated costs of fielding a team of statisticians with the costs of financing a twinning arrangement by Statistics Sweden. It found that the twinning arrangement was almost 30 percent more expensive.[29] But the SIDA evaluation concluded that it was worth paying more to get the added benefits "built into the institutional cooperation model."

BOX 3.1
TWINNING IN SENEGAL

> Senegal's Société Nationale de Télécommunications (SONATEL) has a twinning arrangement with France-Telecom/Auvergne, headquartered in Clermont-Ferrand. France Telecom performs, under contract, several services for SONATEL. It tests the equipment that SONATEL buys, for example. When SONATEL introduces new systems or technologies, the company often calls on its French twin to test it, and to help with programming and implementation. This was the case, for example, with a new mobile radio-telephone system.
>
> SONATEL sends people to the Auvergne to work with France-Telecom. Training occurs at all levels of the work force. The power line foreman, for example, has worked with his counterpart in France.
>
> The relationship is informal. There are no contractual relationships between the parties. Each year SONATEL defines its needs and finances training and other activities out of its own resources.
>
> This arrangement works in large part because France-Telecom has an economic interest in the smooth functioning of SONATEL. If as many as 20 percent of French calls to Senegal fail, the costs to the French system are substantial. Twinning in this case is successful in part because France is a major market for Senegalese telecommunications. If market shares were more dispersed, none of the developed country partners of Senegal would have an incentive to help Senegalese performance.

CONCLUDING REMARKS

We have seen in this chapter that many deficiencies have always been present in technical cooperation delivery systems—from faulty nuts and bolts, as revealed in poor project design and implementation, to excessive reliance on an unsuitable vehicle for capacity building, the resident expert-counterpart arrangement. These problems have been described and deplored for some 15 years, yet they persist.

Most of the proposals for change are based on strong exhortation—urging the parties, especially the donors, to mend their ways. This may work. Aid organizations can do and have done better. But based on past experience not much change can be expected from exhortation alone. Ingrained organizational behavior and vested interests lean the other way.

Efforts to put better technical cooperation projects together, and to implement them more carefully, should certainly not be abandoned. Nonetheless, it seems likely that real reform is more likely to come from changing the basic model of technical cooperation delivery. We, therefore, follow what has become conventional wisdom in recommending drastically reduced reliance on the expert-counterpart model. Its flaws are so deep that even the best-designed repairs cannot be expected to make it a significantly better instrument for capacity building.

That being said, we have to admit some uneasiness about what is at hand to replace the resident expert-counterpart model, which is the foundation of technical cooperation arrangements worldwide. The proposals put forward here—for unabashed gap-filling, short-term coaching technical assistance personnel, greater use of local consultants, and twinning seem sensible and right. Most of them also form part of the new consensus about what is to be done. But except for placement of technical assistance personnel in line posts, experience with the substitute vehicles is limited. We thus have, on the one hand, a traditional instrument that is familiar and comfortable for most of the players but that has been demonstrated to be bad for capacity building. And, on the other hand, we have a set of proposed substitutes that seem better suited for that task but that are untried. It is easy to understand why technical assistance strategies have become so lively an issue in the development community.

NOTES

1. A paper written for a seminar on technical cooperation in Haiti emphasizes the same point. "Technical assistance perhaps plays a role different than that which is claimed: taking account of the past misuse of aid, donors seek greater control, and this is a reason why they need to have senior technical assistance people on the ground. Often their role is mainly to provide financial oversight." Gouvernement Haitien/PNUD, "Quelle assistance technique pour Haiti?" juin 1990.

2. See, among others, Francis Lethem and Lauren Cooper, "Managing Project-Related Technical Assistance: The Lessons of Success," World Bank Staff Working Paper no. 586, 1983; and the Forss report (Kim Forss, J. Carlsen, E. Froyland, T. Sitari, and K. Vilby, "Evaluation of the Effectiveness of Technical Assistance Personnel Financed by the Nordic Countries," 1990).

3. This section is based on Sakiko Fukuda-Parr, "The Expert-Counterpart Model of Technical Cooperation: Debunking the Myths," internal paper, UNDP, New York, 1990.

4. See OECD, DAC, "Principles for New Orientations in Technical Cooperation," Paris, 1991. The World Bank, in its long-term perspective study on Africa (*Sub-Saharan Africa: From Crisis to Sustainable Growth*, 1989) says: "The long-term target must be to reduce technical assistance sharply. The first step is to replace long-term experts with short-term consultants."

5. See the policy report statements for Malawi, Burundi, and Guinea.

6. See comments in Chapter One, and in R. Cassen and Associates, *Does Aid Work?* 1986, p. 197.

7. T. Davis, "Assessment of Institutional Capacity to Effectively Use Technical Cooperation in the Agriculture, Forest, Fishery and National Parks and Wildlife Sectors," July 1990.

8. Government of Guinea Bissau, "Situation et perspectives de l'assistance technique en Guinée Bissau," document de travail présenté à la Conférence de Table Ronde, Genève, juin 1988. These circumstances are also found outside of Africa. In Haiti, for example, three times as many resident experts as professional nationals were employed in agriculture, education, public works, and health agencies in the mid-1980s—310 compared with 102 nationals. Monique Pierre Antoine, "Rapports et interactions de l'Assistance Technique et de l'Administration publique," unpublished note, UNDP, n.d.

9. Similar recommendations are made in earlier World Bank reports.
   - "The borrower should participate in all phases of design, identification of needs, specification of TA delivery models, drafting of terms of reference, etc." (internal report of the Public Sector Management Unit, written in mid-1980s). "Projects should be less sophisticated and more closely tailored to the absorptive capacity of the recipient" (Storrar Report, 1982). "Focus on a few manageable and achievable objectives, and articulate them to leave no room for ambiguity." "Specified outputs and objectives (with benchmarks of measurable performance) should be included in all terms of reference" (reports by R.B. Sunshine Associates, consultants).
   - "Borrowers should establish a TA planning framework within a national manpower plan" (report by R.B. Sunshine Associates, consultants). "Experts should be selected not only on the basis of their technical expertise but also on the basis of their training capability" (Storrar Report, 1982).

10. Michael Stevens, "TA for Institution Building: Some Controversial Thoughts," informal memo, World Bank, 1992.

11. Beatrice Buyck, "The Bank's Use of Technical Assistance for Institutional Development," Working Paper Series, no. 578, World Bank, January 1991.

12. Personal communication from Michael Stevens. Similar circumstances prevail in other countries—Papua New Guinea, for example.

13. OECD, DAC, "Principles for New Orientations . . . ," 1991, par. 62. A World Bank review makes a similar statement: "More thought could be put into the packaging of ID-TA delivery modes, for example by integrating short-term consultants and long-term technical assistants with national staff, and making use of less traditional methods such as twinning. Long-term expatriate advisers should be employed only in exceptional circumstances after alternative methods of delivery,

including the use of local consultants, have been fully explored." Beatrice Buyck, "Technical Assistance as a Delivery Mechanism for Institutional Development: A Review of Issues and Lessons of Bank Experience," Country Economics Department, Public Sector Management and Private Sector Development Division, World Bank, p. iii.

14. A.I.D. has some well-developed administrative arrangements for these purposes, such as Indefinite Quantity Contracts, which are enabling contracts that allow short-term consultants to be hired on short notice and without competitive bidding. Other bilaterals appear to have similar contracting devices, as does UNDP (prequalification of consulting firms and retainer contracts).

15. OECD, DAC, "Principles for New Orientations . . . ," 1991, p. 7.

16. "Changing Patterns in UK Technical Cooperation," note by the U.K. Delegation for the DAC Meeting on Technical Cooperation, November 15-16, 1988.

17. See Marie-H. Birindelli, "Le Cas de la Coopération Technique ACP-CEE: Expériences et Orientations," contribution au colloque organisé par le Centre Européen de Gestion des Politiques de Développement et l'Africa Leadership Forum," Maastricht, 18-20 octobre 1991, p. 7.

18. UNDP, "Implementation of New Dimensions in Technical Cooperation: Report by the Administrator," October 1976, prepared for the twenty-third session of the Governing Council. The World Bank defines twinning as "a professional relationship between an operating entity in a developing country and a similar but more mature organization in another part of the world." L. Cooper, "The Twinning of Institutions," World Bank Technical Paper no. 23, 1984, p. 2.

19. For instance, in 1965, the British Post Office established the British Postal Consultancy Service as a division of professional consultants that drew upon the wide experience of the British postal system. Similarly, in 1969 British Rail set up the United Kingdom Railway Advisory Service, now known as Transmark, to provide advice on passenger and freight railway systems. British Electricity International Limited was established by the United Kingdom's Electricity Supply Industry; it supports linkages such as the twinning arrangement between a power station in Hwange, Zimbabwe, and a technically similar coal-fired power station in the United Kingdom.

20. Cooper, 1984, p. 6.

21. See K.S. Sayed and Warren Baum, "The Consulting Profession in Developing Countries; A Strategy for Development," World Bank WPS 733, July 1991, pp. iv. and 50; and Albert H. Barclay Jr., "Successful Adaptation Means Adopting Twins," in *Developing Alternatives*, publication of Development Alternatives, Inc., Bethesda, Md., July 1991.

22. Other sectors include law and journalism. See "Law Initiative Offers Aid to Third World," *The Financial Times*, August 29, 1989, p. 5; and "International Press Initiative," UPI, March 3, 1981.

23. World Bank, "Bank-Financed Technical Assistance Activities in Sub-Saharan Africa," Washington, D.C., May 2, 1989, p. 28 and Annex I.

24. Cooper, 1984, Annex I.

25. Bertil Oden, Bertil Egero, and Halvard Lesteberg, "Assessment of Institutional Cooperation as a Form of Aid Based on Statistics Sweden's Cooperation with Central Statistics Office, Zimbabwe and Bureau of Statistics, Tanzania," June 16, 1986, p. 13.

26. Participants in the World Bank Second Llilongwe Water Supply Project in Malawi operate in this manner. Employees of the Llilongwe Water Board meet regularly with employees of the South Staffordshire Water Company (United Kingdom) to review progress and then modify or intensify efforts based on their assessment.

27. Buyck (1989), Annex 4, describes a twinning arrangement between Ghana's state gold mining corporation and a consortium of three Canadian mining companies. Mainly as a result of the inexperience of the Canadian partners in dealing with the sociopolitical and economic milieus of developing countries, the project generated much conflict and required frequent arbitration by its financier, the World Bank. It failed and was terminated after three years.

28. Interview in Guatemala by Ann McDermott, staff economist of Development Alternatives, Inc.

29. Oden, Egero, and Lesteberg, 1986, p. 13.

4

STRENGTHENING RECIPIENT MANAGEMENT

Sub-Saharan Africa is the most heavily aided region in the world. During the decade 1980-1989, official development assistance averaged more than 11 percent of GNP in the region, 10 to 20 times more than the other developing regions.[1] In 1989, 12 African countries received official development assistance amounting to 15 percent or more of their GNP, and in 22 countries (about half of those for which data exist), the aid to GNP ratio exceeded 10 percent. In five additional countries, official development assistance was 5 to 10 percent of GNP.[2] As noted earlier, much of this aid—almost a quarter in 1989—is in the form of technical cooperation.

Much of the technical cooperation management problem is embedded in these numbers. All heavily aided countries have great difficulty in managing the foreign assistance they receive. Aid inflows typically involve a large number—hundreds—of separately negotiated and implemented projects sponsored by many different donors, each of which has its own set of rules and regulations on everything from accounting practices to procurement and evaluation requirements. At the same time, most of these countries fall in the category of least developed, and are characterized by weak administrative structures.

For these countries, the task of dealing with a large number of projects and a multiplicity of donors is overwhelming.[3]

The most fundamental general problem of technical cooperation management is that it is managed too little by African governments and too much by aid donors. This is the meaning of the universal criticism that technical cooperation in Africa is supply driven. On the macroeconomic level, knowledge and control are spotty: recipient governments know only roughly how much technical cooperation they receive, and they exercise little influence over its ebbs and flows. Governments have little say in the way donors allocate their assistance between technical cooperation and other forms of aid.

Management of technical cooperation at the micro or project level also unfolds with small government input. The initiative for conception, design, implementation, and monitoring of technical cooperation projects comes mainly from the suppliers. The few inputs made by recipient governments come almost entirely from the user agencies; central screening and priority-setting bodies (planning and finance ministries) impose only sporadic and highly imperfect coordination and control.

Donor dominance, or what has been called the unequal relationship, is now universally deplored, and all parties recognize the need to transfer into national hands more responsibility for technical cooperation management.[4] Without it the effectiveness of technical cooperation will continue to be eroded by lack of local ownership, unproductive or low priority use of technical cooperation, and weak commitment to sustain technical cooperation projects. National management is not by itself sufficient to eliminate these problems, but it is necessary.

This chapter analyses the nature and sources of technical cooperation management problems in Sub-Saharan Africa and examines the measures that have been proposed to increase local control. The chapter has four parts. The first describes the general decision-making environment for technical cooperation. Because the problems of effective technical cooperation management are extreme manifestations of the general problem of aid management, typical decision-making arrangements for public investment projects are briefly outlined. This provides context for a discussion of technical cooperation decision making.

The second part sets out the pattern of donor dominance at all stages of the project cycle and at the macro level. The third indicates the costs and inconveniences of the present arrangements for technical cooperation management and suggests some of the reasons for their persistence despite much criticism. The final section addresses the question of reform tactics and strategies—what has to be done to strengthen local management of technical cooperation acquisition and use.

THE CONTEXT: DIFFUSE PUBLIC SECTOR DECISION MAKING

Because of the way technical cooperation decisions are usually made, recipient countries have little opportunity to define their needs or to choose what kind of technical help they want. One basic reason for this state of affairs is that public sector decision-making processes in most African countries are diffuse and disorderly. The point is best illustrated by considering typical decision processes for public investment projects.

Capital Investment Projects

Two sets of actors, one domestic and the other external, determine the number and nature of capital investment or development projects in most African countries. The domestic actors fall into two groups: the spending agencies that are responsible for the implementation of projects, and the central screening agencies (usually planning ministries and finance ministries) that are responsible for setting priorities in public expenditures and public policy.[5] Interacting with the domestic actors is a set of external players—bilateral and multilateral aid donors and private investors or other agents (nongovernmental organizations, for example)—seeking to operate in the country in question.

Most projects are identified and elaborated by some combination of donor and host government action. The typical investment project in Sub-Saharan Africa arises from a donor initiative, with an operating agency usually playing a collaborative but passive role. The project may be carefully prepared and closely linked to national priorities. Or it may be shoddily put together or low in priority. In all cases, once prepared, written up, and put on the table, it is likely to become part of the public investment portfolio.

This is so because internal project screening procedures are porous and often allow low-priority projects to pass through the evaluation net. The implementing agencies do little internal vetting; their interest is to have more rather than fewer projects. It is at the level of the finance and planning ministries that the critical assessments should take place. But finance ministries are concerned primarily with holding down aggregate expenditures; they are rarely able to control the quality of expenditure and are sometimes not much interested in it. In practice, they therefore do little real evaluation of the soundness or priority of proposed development projects.

The job of screening projects and assuring adherence to priorities thus falls to planning agencies. But these agencies are with rare exceptions weak. They have small staffs, small nonsalary budgets, and limited access to technical ministries. Absorption in other tasks, such as writing medium-term development plans or long-term perspective studies, leaves them little time for in-depth evaluation of projects. Most in any case lack the sectoral competence that is essential for effective project screening.

Real progress was made in the 1980s in imposing greater rationality and discipline in investment programming. As part of the adjustment process, investment budgets have become more transparent; many Sub-Saharan Africa countries have moved toward comprehensive budgeting, making foreign-financed investment expenditures explicit in the budget. Multiyear rolling public investment programmes have been widely introduced. Repeated decrees and regulations requiring approval of all investment projects by the ministries of planning and finance have led to greater discipline and more economically rational project choice.

This progress in investment programming, however, is recent and uneven; in some countries it remains shallow. Adherence to the approval process is still imperfect, and its impact on raising the quality of investment decisions is not always clear. Planners and screeners are frequently presented with faits accomplis, projects that have been tightly packaged by donors and implementing agencies by the time they are presented for formal vetting. And the rationalization of project selection—close and systematic project screening and prioritizing by planning ministries or budget bureaus—still has a long way to go in most countries of the region.

It is not surprising that investment programming processes are heavily donor driven in Sub-Saharan Africa. Well over half the capital investment in most Sub-Saharan Africa countries is aid financed; in many countries the figure is closer to 80 percent. Although this phenomenon of external dependence for investment financing has become particularly striking in recent years, it has been true for several decades. Donor agencies identify and elaborate many, probably most, public investments. They have the determining voice in deciding the scale of projects, the technology used, the size and mix of staffing, and how the projects should be implemented and monitored. Increasingly, donor agencies pay some of the local costs associated with "their" projects.

Recipient government management and control of capital projects thus remain partial and tenuous. But the nature of the problem has been clearly identified, its importance is recognized, institutional reforms have been put in place in many countries, and improvement is widely observable.

Technical Cooperation Projects

The management and coordination problem is more acute for technical cooperation than for capital projects. Free-standing technical cooperation projects tend to be small in size, but numerous; they may number in the hundreds even in the smaller African countries. Investment projects, most of which have technical cooperation components, also number in the hundreds.

The number of suppliers and projects is thus staggeringly large in most of Sub-Saharan Africa. Even a decade ago, according to one report, 82 multilateral and bilateral donors offered significant amounts of development assistance to Sub-Saharan Africa.[6] In the early 1980s, for example, Malawi had 188 projects from 50 different donors and its Ministry of Agriculture alone was responsible for 44 donor-financed projects; Zambia had 614 projects from 69 donors, with the Ministry of Agriculture and Water Development accounting for 120 projects; one of the smallest countries in Africa—Lesotho—had 321 projects from 61 donors. Recent studies under NaTCAP and other auspices indicate that these numbers have not declined over the 1980s. Senegal's Three-Year Public Investment Programmes in the late 1980s had over 400 projects, without counting most free-standing technical cooperation projects. Some 1,400 were counted in Burundi.

The management requirements of the large number of donors and their great diversity are immense. Each donor has its own charter that determines the kind of development activities it can pursue. Each has its own project preparation requirements, its own project implementation procedures, and its own reporting requirements; and each is responsible to its own governing council.

African governments are asked to carry out activities and provide information that respond satisfactorily to the interests of each donor. To do so involves countless meetings with donor representatives —officials, experts, consultants, and suppliers—at each phase of the project cycle. For each project, governments must provide information and counterparts for identification, feasibility, and design teams. Project documents must be examined and project agreements negotiated. During implementation, governments are expected to provide management and counterpart personnel, an operating budget, and logistic support. They are required to manage a myriad of separate project accounts in compliance with complex donor procedures. Governments must also generate a blizzard of progress reports—often on a monthly basis—in the formats laid down by each donor.

In many African countries, donor demands have stretched management and administrative capacities to the breaking point. Under the circumstances, it is not surprising that governments are frequently unable to manage inputs effectively. Nor is it surprising that the donors, who are constantly looking over their shoulders at the potential reactions of their governing councils, have become frustrated at what they perceive as recipient governments' management shortcomings and have, as a result, increased the intensity of their own development management activities.

DONOR DOMINANCE OF TECHNICAL COOPERATION MANAGEMENT

More even than capital projects, technical cooperation is widely perceived to be donor driven—its management, that is, is more in the hands of donors than of recipient governments or organizations.[7] Its donor-driven nature is manifested in several ways.

First, ideas for technical cooperation projects frequently originate with donors. In some cases, projects are identified solely by recipient governments. This is rare, however. More often, identification is a joint product resulting from early collaboration between donors and operating agencies in the recipient country. But in most cases ideas are hatched in resident missions or donor capitals and are then sold to recipient governments. Governments are not always convinced of the need for, or intrinsic merits of, technical cooperation projects, but are persuaded to accept them because they are presented or perceived as conditions for obtaining capital or budget assistance; the financial and opportunity costs of accepting them seem low; and host country managers believe that the resources of the technical cooperation project may be useful, although in ways that are different from those perceived by the donors.[8]

Second, most technical cooperation projects are prepared or designed in large part by the staff of, or consultants to, donor agencies rather than by recipient governments. Donors prefer to use their own staff or consultants for this purpose because (1) they are a known entity; (2) they are familiar with the formal and informal requirements and procedures of the project design process; (3) it is easier, culturally and linguistically, to work with people from one's own organization or own country; and (4) the design will be delivered in a format and on a schedule acceptable to the donor.[9] Recipients, for their part, go along with this because they may lack the staff or the resources to design the project on their own or simply because the design of projects by donors has been the practice for so long it has become accepted as normal.

Third, donors or donor personnel often dominate the implementation of technical cooperation projects. Each donor has a specific legal arrangement defining the responsibilities of the donor and the recipient about the control of project resources. These are frequently referred to as "executing arrangements" in the U.N. system, and can vary widely.

At one extreme, government has rights of signature on the use of project funds, controls the use of project equipment such as vehicles, and exercises supervisory control over the performance of technical assistance personnel. Donors retain strong oversight and audit mechanisms. This is the case for World Bank projects that are loan financed. "Government execution" in the U.N. system is a similar arrangement; it approaches full management by government.

At the other extreme, government has little control over the use of project resources. Donors provide these in kind through direct procurement and approve each and every expenditure. Frequently used instruments of donor control of implementation include (1) the assignment of a chief of party with executive authority over project materials (automobiles, operating funds, and so forth); and (2) creation of project management units that have the effect of removing technical cooperation resources from the line authority of established recipient government organizations. These are more common for capital projects than for technical cooperation, and nowadays project units are often run by national directors.

Finally, midproject or ex post evaluations of technical cooperation projects are most often carried out by donor staff or consultants. The terms of reference for these evaluations, moreover, are usually prepared by the donors to comply with the regulations and interests of their governing bodies rather than to meet the decision-making needs of recipient government managers.

COSTS OF THE PRESENT ARRANGEMENTS AND WHY THEY PERSIST

Diffuse decision making and donor management of technical cooperation have had several unhappy consequences.

- Technical cooperation activities are subjected to little internal scrutiny in terms of need, impact, and priority. Free-standing technical cooperation projects are rarely vetted in public expenditure programming exercises and, except in a few countries, are neglected in public investment programme documents. Nor are many questions raised during investment programming exercises about the technical cooperation components of capital projects. Their adherence to national priorities is therefore accidental or haphazard: they conform to donor rather than local priorities. Although these undoubtedly overlap, the fact that supplying agencies shape to so great an extent the volume and type of technical cooperation brings high risks of divergence from national priorities.
- Because technical cooperation activities are subjected to less intensive economic analysis than capital investments and because these activities are not considered a macroeconomic variable of importance, there has been a common tendency to have too many projects—that is, too many of low priority. Possible beneficial trade-offs between technical cooperation and other forms of aid have been neglected. The inability of host governments to control and coordinate technical cooperation operations leads to a proliferation of overlapping, sometimes competing, technical cooperation projects, which contributes to low productivity.
- Donor management is seen as donor ownership. This dilutes recipient interest in and commitment to technical cooperation and leads to poor use of technical assistance personnel. Effectiveness and sustainability are undermined.

- That management in general and supervision, monitoring, and evaluation in particular make up a process internal to the donor-supplier limits local involvement and local benefits from feedback, thereby reducing institutional learning. Donors pay for the evaluations, for example, which are directed at their concerns much more than those of local agencies.
- Technical cooperation personnel are often placed in the difficult position of reporting to and being managed by at least two bosses—the donor agency or consulting firm that hired the consultants and the agency for which they work on the ground. Resulting problems of uncertain lines of authority and divided institutional loyalties among technical cooperation staff can complicate the management of the public sector. If push comes to shove, it is usually the salary-paying donor organization that has first claim on loyalties.
- A new kind of dependency has emerged in many countries of the region. Because the large volume and diversity of technical cooperation resources overwhelms existing national management capacities, donors respond by gradually increasing their own control over the resources they provide. Meanwhile, local officials—aware of their limited ability to change the flawed technical cooperation relationship—adopt a strategy of passive acceptance, which adds to other forces making for deepened dependency.

These deficiencies have long been recognized and widely commented on. Why do they persist and even grow more burdensome? Five principal reasons can be identified.

First, weak management of technical cooperation is part of the general administrative and institutional weakness that characterizes low income countries generally. A striking paradox prevails for heavily aided countries: they face the most demanding challenges of aid administration (and development management generally) and are least equipped, in institutional capacity, to meet these challenges. Weak capacity at all stages of the project cycle and in macroeconomic management is endemic and deep rooted in most low-income countries. The lack of these capacities is indeed a characteristic of underdevelopment. Given the general institutional constraints, it is hardly surprising that technical cooperation is poorly managed.

Second, the administrative environment deteriorated in the 1980s. Government wage and salary policies and declining nonsalary budget resources reduced public sector capacity by eroding work incentives for skilled people and by providing fewer materials for them to work with. Donors responded to this situation by imposing more controls on their technical cooperation and other aid, and more technical assistance personnel to provide oversight or gatekeeping functions. A vicious circle was established: as recipient government capacity to manage declined, donors took on more management responsibility themselves.

Third, progress in transferring responsibility to recipients may have been hindered by too-narrow perceptions about the sources of administrative deficiencies in recipient countries. Many observers, including some donor representatives, stress underlying factors such as lack of political will (as expressed in inadequate and inefficient budget allocations and acceptance of corruption); institutional weaknesses such as nepotism and inability to fire people for poor performance; poor use of existing staff (little delegation of responsibility and few staff meetings); and poorly developed procedures in critical areas such as project preparation and evaluation, financial monitoring, policy analysis, and debt management.

Few recipient agencies would deny the existence of these problems. But in the aid dialogue they tend to be downplayed. Poor performance is attributed mainly to insufficient or underqualified staff and to insufficient equipment or operating funds. Management problems, in other words, are seen in terms of human resource and budgetary shortcomings rather than as rooted in institutional deficiencies. Thus institutions do not need to be restructured or privatized or made more accountable and transparent in their procedures. They just need more better-trained and better-paid personnel. They need money (via budget support) and technical cooperation for a little gap-filling technical assistance personnel and for training.[10]

That everyday administrative problems have been perceived so much this way has probably diverted attention from the institutional prerequisites to strengthened administration. That donor agencies have themselves neglected institutional analysis, or done it badly, is also pertinent.[11] One reason may be that behavior patterns, which are rooted in social values, are taken as given by project makers and implementers. Behavior patterns are seen as part of the cultural landscape, so general and so profoundly rooted as to be beyond the reach of simple administrative reformers. To change these behaviors is the meat and potatoes of institutional development, but is an indirect and long-term process.[12]

Not only has the definition of the problem deflected attention from institutional issues, but the extensive use of expatriate technical assistance personnel also has not facilitated problem solving. The types of adaptations required are not ones that are easily engineered by expatriate technical assistance who usually have little understanding of recipient country culture.

A fourth factor has to do with recipient country incentives. Many recipient country officials see technical assistance personnel (the largest component of technical cooperation) as a free, or almost free, good. There is, therefore, little incentive for recipient agencies to want to manage it. The components of technical assistance management—preparation of scopes of work, advertising, interviewing and selection, negotiating, contract preparation, travel and transportation, payroll, benefit administration (often including provision of housing and payment of utilities), work supervision, and reviews of performance—are administratively burdensome. At present, most of these management tasks are performed by the donors that provide the technical assistance, and many recipients, short of staff and otherwise occupied, are happy to leave it that way.[13]

Finally, donors have little incentive to transfer to recipients management responsibility for their technical cooperation resources. Donors are, first and foremost, responsible to their governing councils for the effective and efficient use of their resources. The more they transfer control over these resources to recipient governments the less able they are to affect how the resources are used. In international meetings and in many documents, donor spokesmen appear to agree that technical cooperation is too donor driven. But at the operating level, few donor representatives are willing to relinquish control over such elements of technical cooperation management as programme planning, project design, expenditure tracking and accounting, and oversight of the compliance with regulations.

The reasons are obvious. With regard to programme planning, the governing bodies of all aid agencies establish priorities for the use of their resources—for example, natural resources management, governance and human rights, poverty, and women in development. Local aid agency missions have to adhere to these priorities, which may not always accord with host government priorities. Donor agencies also sometimes have strong views on how technical cooperation projects are to be implemented; they may prefer private sector channels, whereas governments may find public sector implementation more appropriate. And user agencies may use technical assistance personnel in ways that conflict with their terms of reference, or change the mix of inputs in the technical cooperation package—overspending on equipment and underspending on training, for example. In many countries, misappropriation or corruption is feared.

THE NEED FOR REFORM

Notwithstanding these constraints, a general recognition has emerged that the unequal relationship between donors and recipients with respect to technical cooperation management is a fundamental barrier to effective capacity building and that changes are essential. Improvements are needed by both donors and recipients.

Donor Coordination

Until recently, reform proposals focused mainly on the donor side—the frequent exhortations to improve technical cooperation project design and delivery, for example. And most of the improvement in aid coordination (including the coordination of technical cooperation) that has occurred has been through donor initiatives, such as informal exchange of information at the national level, sectoral cofinancing meetings organized mainly by the World Bank, and formal Round

Tables and Consultative Group meetings organized by UNDP and the World Bank. This has probably reduced duplication and led to better-informed donor programming, and some better-organized sectoral aid allocations.[14]

These donor-led and donor-sustained initiatives can make the technical cooperation process more orderly, but they cannot be expected to have more than marginal impact. The most effective donor coordination takes place at country-level meetings, usually informal or sectorally focused. Although the prevalence and workings of these informal coordination vehicles have been little studied, casual observation suggests that they often spend a lot of time airing donor grievances, that they are not usually set up to allow in-depth technical discussion, and that changes in donor staff make continuity difficult. Round Tables and Consultative Group meetings provide for much exchange of information, especially in the corridors. But their contribution to better aid (especially technical cooperation) coordination is limited by numerous constraints: their multiple objectives, the high level of generality of their discussions, the priority they give to financial gap-closing, and the formalism of their procedures.[15]

Donors can take specific actions to improve aid coordination in general and technical cooperation coordination in particular. They can make coordination a separate item at Round Tables and Consultative Groups; strengthen resident aid missions and country-level coordinating arrangements; give more attention to their mix of technical cooperation, policy lending, and sectoral and project loans, and to the complexity of their programmes; and harmonize their procedural requirements, especially for projects that are cofinanced.[16]

A solution proposed by several observers is that donors simply limit the number of projects they finance, including technical cooperation projects. Thus, the Maastricht paper on technical cooperation cites with approval a major recommendation in the 1986 report, *Does Aid Work*:

> The best solution would be for governments to set priorities for institution building; for governments and aid agencies to design technical cooperation programmes around these priorities; and

for donors to combine in a limited number of projects that would drastically reduce the aid management burden on the government. Such action would require from donors rather more than the 'harmonization' of procedures that has been discussed within the DAC.[17]

Such proposals have found little support in the donor community. One reason is that they appear to call for a retreat in the face of growing needs. This is not acceptable to donors. In fact, aid projects have multiplied in Sub-Saharan Africa, and—even more burdensome administratively—numbers of donors have also increased.

But there are more basic reasons why recommendations to integrate and better coordinate aid inflows have proven infeasible. Almost all observers of aid coordination problems note that although donors support coordination in the abstract most are unwilling or unable to be coordinated. Each has different administrative styles, bureaucratic requirements, political objectives, and programme agendas. The Maastricht study summarizes a consensus opinion as follows:

> Problems of donor coordination have shown [sic] to be more intractable than commonly assumed, arising as they do from deep-seated organizational and attitudinal obstacles within the donor community, in particular the lack of incentives to work with other donors and the obsession with achieving expenditure targets.[18]

More Effective National Management

Some amelioration of technical cooperation management is certainly achievable by donor action alone—for example, better coordination and improved project design and implementation. But, for reasons noted above, gains from that direction are likely to be limited. In any event, more effective management by governments is indispensable. Without it, donor-driven improvements are unsustainable. And it is from better African government management that most of the benefits will arise: a greater sense of ownership of technical cooperation projects, greater certainty that priority needs are being addressed by technical cooperation, greater rationality in establishing the balance between technical cooperation and other forms of aid, better quality technical assistance personnel and greater commitment to its effective use, closer attention to cost issues, and greater prospects for sustainability.

At a general level, it is easy to see what has to be done. First, basic information gaps have to be filled. The authorities in many countries do not have up-to-date information on numbers of technical cooperation projects; their breakdown between technical assistance personnel, equipment, training, and operating costs; their sectors and national origins; and their terms.

It is telling that information on technical cooperation, even on its size in terms of spending and numbers of technical assistance personnel, is sparse in most countries.[19] Those considerable technical cooperation expenditures that do not pass through the local budget process are often poorly known—payments to resident technical assistance personnel, for example, and off-shore procurement. Except in countries that have introduced comprehensive budgeting, budget documents are usually silent on these expenditures. Public expenditures are therefore often underestimated by significant amounts, at least in countries without comprehensive budgets.

Nor is it clear in most country economic data whether and how the value of technical cooperation projects is included in the balance of payments numbers or in the national accounts more generally (see Box 4.1). Until recently, project-related technical cooperation has rarely been broken out for separate statistical or analytic treatment.

Second, some policy consensus should be achieved regarding the role of technical cooperation—how much and what kind is needed, how its effectiveness can be increased, how it ties in to national objectives. Also, individual project proposals should be evaluated for their soundness and priority and provision made in the state budget for their financing. And the organizational and procedural mechanisms that perform these functions have to be put in place—a technical co-

BOX 4.1

ACCOUNTING FOR TECHNICAL COOPERATION

Expenditures on technical cooperation are a major macroeconomic variable in Sub-Saharan Africa. They represent a quarter of net official development assistance flows, and for the poorest African countries, they amount to about a third of export earnings, a quarter of government revenues, and almost 5 percent of GDP. Yet their treatment in national accounts, balances of payments, and central government budgets is frequently unsystematic and uncertain.

In many instances they are not counted at all. This is evident from the observed differences between capital flows as recorded in donor statistical sources and inflows reported by developing countries. The balance of payments statistics of recipients invariably give smaller numbers than are found in the DAC data. The largest source of these differences appears to be technical assistance grants.

How do national income, balance of payments, and budgetary accounts treat technical cooperation expenditures? Practice varies greatly between countries. But the prevailing rules are set down in the three "bibles" of national accountants: the United Nations' Systems of National Accounts (SNA); and the International Monetary Fund's Manual of Balance of Payments (MBP), and Government Finance Statistics (GFS). The following discussion of expenditure classification is based on the guidelines set down in these sources and their companion documents. It should be noted that governments do not always follow these guidelines.

Take a typical institution-building technical cooperation project, consisting of long-term resident expatriate experts, short-term consultants, training (in-country and abroad), local employees, and some equipment.

- **Long-term expatriate resident experts** are treated as residents of the country they are working in, if they are present more than one year. Their compensation is counted as part of domestic value added, hence part of GDP, national income, and national expenditure. Their financing, assuming it is a grant from a bilateral donor, is recorded as a transfer from abroad. The MBP classes these salaries as unrequited official transfers and, in the GFS, technical assistance is a grant in kind and is only recorded *pour mémoire*. In fact, the IMF gives no guidance in its GFS handbooks on how to treat these expenditures, so each government follows its own practice. In many cases, technical assistance and expenditures of this kind and others are not included at all in official budget documents, because they are mainly grants with disbursements often made in donor capitals.

BOX 4.1 — Continued

- **Short-term consultants** are considered nonresidents, and noncontributors to domestic value added. Their wages are counted as payments for imported services or foreign factor income.
- **Trainees** sent abroad are considered residents of their country of origin, no matter how long they study abroad. In the national accounts, this is regarded as consumption of foreign goods and services. The MBP counts these expenditures as payments for imported services (travel).
- **Salaries of national staff** employed on a bilateral-financed project are recorded as either wages or consumption of services. Their payments do not appear in the balance of payments. (However, payments to local staff employed by international organizations do count in the balance of payments because salaries paid by these organizations are considered as factor revenues received from abroad.)
- **Capital equipment** is counted as investment and classified under imports of goods and services.

Analytic inadequacies and data deficiencies regarding technical cooperation persist, and require attention. Some important expenditure items are vaguely and variably treated in current practice. Local government cash and in-kind contributions to technical cooperation projects are probably the most important example.

Basic data remain sparse, even on numbers of technical assistants at the country level, although more is being generated now thanks to such information-generating efforts as the NaTCAPs. But little remains known about such significant data as that related to household expenditure patterns of technical assistance personnel, especially savings and import propensities.

Some methodological issues remain unresolved. Presently, for example, long-term resident technical assistance personnel are counted as domestic factors of production, contributing to domestic value added. But intermittent consultants, returning for periodic visits on the same consultancy, are counted as foreign factors of production. Thus, a shift from resident expatriate experts to intermittent consultants, under present definitions, diminishes GDP and national income.

All of this suggests a need for more intensive efforts, both to generate more and better data and to give more methodical treatment to technical cooperation in the main macroeconomic aggregates.

Source: This box is based on the analysis in an unpublished paper by Maria Jose Laranjeiro, "Intégration des Projets de Coopération Technique dans le Processus Budgétaire et de Planification Nationale," UNDP, September 1990.

operation coordinating unit, for example, and a technical cooperation programming instrument.

More concretely, effective technical cooperation management has three sets of requirements. First, the core economic agencies (ministries of finance and planning) have to set down broad policies and define priorities with respect to global public sector financial requirements and the place of technical cooperation financing in the macroeconomic picture. In collaboration with operating agencies, these core economic units have to define sectoral and subsectoral objectives and associated manpower needs.

The core economic agencies also have to establish guidelines (ceilings) for technical cooperation expenditure and for sectoral allocations. Policies regarding use of different types of technical cooperation inputs are also required—the priority given to resident versus short-term expatriate personnel, for example, and the relative weight to be given to national experts and consultants.

The regulatory framework for use of technical cooperation has to be specified, including:
- Methodologies and guidelines for design of technical cooperation projects and criteria for project selection;
- Procedures for recruitment of expatriates and national personnel, policies on pay and other conditions of employment for all technical assistance personnel, and rules on incentive payments to civil servants;
- Procedures for selection of trainees, and the establishment of a framework for technical cooperation-related training that is embedded in some overall strategy for work-force development; and
- Rules on how technical cooperation projects should be treated in the budget process and especially the way they relate to the public sector investment programme.

As part of core agency responsibility for determining aggregate levels and sectoral allocations of public expenditures, the agencies should evaluate projects that use technical cooperation for their suitability and priority, just as they evaluate capital investment projects. They have to ensure the coherence of technical cooperation proposals—see that they are in line with social and economic development objectives and consistent with each other. They also have responsibility for monitoring of financial implementation, notably by checking disbursements against commitments. This is a demanding but essential task, because often technical cooperation not only involves a large volume of resources, but most of those resources also come on grant terms and thus escape other tracking systems, such as those for debt management. And much technical cooperation expenditure is disbursed externally (for example, payments to resident expatriates and foreign fellowships and other training costs), making it especially difficult to know rates of implementation and therefore complicating forward planning.

Second, much stronger technical cooperation management capacity is needed at the level of the operating agencies of government. These are the main users of technical cooperation and are, in principle, responsible for project identification and design (as well as implementation)—elements of the project cycle that are now so often donor dominated. Given the weakness in analytic and planning capacity typical of these agencies, much effort will frequently be demanded to bring them up to required performance levels. The agencies need to develop the capacity to define technical cooperation project objectives in terms of outputs, to define needs in terms of inputs, to cost all inputs accurately, to justify projects in terms of well-defined benefits, and to confront sustainability issues. At some stage, they should be able to weave individual projects into sectoral or subsectoral programmes.

More specifically, for user agencies to take full charge of implementation, they will have to define the specific types of technical cooperation desired, prepare terms of reference for technical assistance personnel, recruit the personnel, specify and procure equipment, define types of training needed and specify who should do the training, select trainees, and evaluate training efficacy. They have to manage the project's material resources such as supplies, vehicles, and office equipment, and do on-line supervision—approving work programmes, for example, and generally overseeing technical assistance personnel activities and evaluating performance.

Finally, to perform these tasks of managing technical cooperation, recipient governments have to create or strengthen certain organizations and procedures.

- A technical cooperation management unit has to be created or reinforced that will have general responsibility for technical cooperation programming and coordination.
- An interministerial policy-making committee is usually necessary to oversee technical cooperation policy at the macroeconomic level (the size of the total programme, for example), to provide other policy guidance, and to give political approval of sectoral priorities and programmes proposed by the technical cooperation management unit.
- An ongoing management information system has to be put in place for regular generation of data on number of personnel, number and type of projects, and monitoring of disbursements.
- The priority list of technical cooperation projects approved for financing has to be programmed in some way within the general public expenditure process.

GETTING FROM HERE TO THERE: PRESENT AND PROPOSED APPROACHES

The general target is clear: to transfer much greater technical cooperation management responsibility to local hands. The managerial inputs needed for improved management are also well understood, as shown above. The big question is how to bring about the needed changes. Three main avenues are being followed or recommended: the outright empowerment of recipient governments by donor hand-overs of authority at the project level, reinforcement of government capacity to administer technical cooperation by NaTCAPS, and a shift from project to programme approaches.

Voluntary Donor Transfer of Managerial Authority

Several bilateral donors are trying to hand over responsibility for management of aid projects, including technical cooperation projects.

These experiments take various forms and names. One that is widely cited is "co-management," under which recipients are made responsible for specific aspects of project management. Thus, under some agreements between the Belgian government and the governments of Burundi and Rwanda, recipients are supposed to identify and propose new projects, including technical cooperation requirements. The donor is to undertake only ex post control, although with some overall responsibility in financial management of the project.[20]

The Nordics, especially SIDA and the Norwegian aid agency, have announced the intention to push further along these lines. They would convert the donor role principally to that of financing agency, with strengthened analytic and evaluation capacities. The idea is that recipient governments would take over all steps in the project cycle, with donors doing ex ante and ex post project evaluation and perhaps playing some role in financial management. The nature and volume of technical cooperation would no longer be decided by the financiers. Donors would presumably have little or no role in determining the makeup of technical cooperation; the choice, for example, between use of local and expatriate technical assistance personnel would be for local authorities to make.

The move to greater direct control by recipients is a subject of debate in many donor administrations. Experience is too patchy and too recent to permit judgments about effectiveness of these new initiatives; indeed, not enough is known about how much real transfer of authority has taken place. It is known that the Belgian effort at *cogestion* has given rise to serious implementation difficulties in the Belgian administration. And UNDP has mandated "national execution," but lack of preparedness on the part of governments has been an enormous obstacle.

NaTCAP

This approach to stronger technical cooperation management was launched in 1986 under UNDP sponsorship. It has now become the main vehicle for technical cooperation management reform; by 1991, the NaTCAP concept had been adopted by 30 African governments.[21]

NaTCAP aims at introducing the processes and building the organizational capacities necessary for effective technical cooperation management by recipient governments. Although initial methodological help and financing came from UNDP, and financial support continues, NaTCAP is government led. And although some observers have expressed doubts about the extent of internalization of the process, they acknowledge that NaTCAP has acquired more local ownership than any other aid reform effort.[22] Certain basic tools will come out of the NaTCAP process that will allow improved use of technical cooperation resources. These are not meant to be cookie-cutters, the same for every country, but are general instruments to be adapted to the specific needs of each country.

The first tool is the development of a management information system that would allow governments to collect and process reliable data on current and proposed technical cooperation projects and programmes. Diagnostic studies aimed at assessing technical cooperation management provide a second set of analytic tools. These depend on country circumstances, but focus normally on such matters as the perception of local project managers about the effectiveness of technical assistance personnel; the methods and procedures used to identify, recruit, and deploy technical assistance personnel; the manpower and training needs of the public service; the role and effectiveness of volunteers and nongovernmental organizations; the availability of local consultants; and the feasibility of repatriating trained nationals.

These studies and the information from the technical cooperation database provide material for a policy paper, the Technical Cooperation Policy Framework Paper, which sets out consensus government positions on such key issues as desirable distribution of aid inflows among, for example, technical cooperation, development projects, nonproject (policy) loans, and food aid. The drafting of the TCPFP is intended not only to draw attention to neglected policy issues, but also to be a consensus-building exercise. The end product also should be useful for distribution at Round Tables and Consultative Group meetings.

A final element is a technical cooperation programme, which sets out priority technical cooperation projects in a sectoral framework and programme needs for technical cooperation resources over a specific period of time. The technical cooperation programme can be a statement of sectoral priorities with a limited list of big or important planned technical cooperation projects. Or it can be more comprehensive—the technical cooperation equivalent of the now-widespread public investment programmes—and involve a full list of priority projects, arranged by sector and consistent with sectoral objectives, fully costed, phased, and integrated into the formal budget process.

In addition to these outputs of information and analysis, the NaTCAP process focuses on institutional requirements—strengthening organizations and procedures needed for better technical cooperation decision making and better implementation of approved projects. The notion of a "focal point" has come to the fore here—a strong unit in a core economic agency that can be given responsibility for coordinating technical cooperation. A political coordinating committee is also frequently recommended. And almost all NaTCAP-related studies have identified weak operating agency capacity for planning and for project preparation capacity as strategic constraints.

The NaTCAP process is recent and still evolving. It required some push from the outside to launch, and UNDP financial support and other encouragement are still needed in most countries where it is under way. Evaluations, both internal and external, point out that, although the diagnostic-analytic aspects have moved ahead quickly and often effectively, operational follow-on has been slow and uneven.[23] Moving from the policy framework into technical cooperation programming has so far occurred in only a few countries. And the idea of creating or reinforcing an organizational focal point to manage technical cooperation within the public sector has proved difficult to implement in at least some cases.[24]

Nonetheless, NaTCAP's achievements are substantial. More policy-relevant information about technical cooperation has been generated under its auspices than ever existed before. Country-focused databases have been created in many countries, and numerous policy inquiries undertaken. The level of awareness of technical cooperation policy problems has been raised significantly, and progress toward internal coordination has ensued, in part from the efforts to derive consensus positions for incorporation in the TCPFP. The way to begin serious programming exercises has been opened by successful innovators in Burundi, Guinea, Guinea-Bissau, Mozambique, and Swaziland.

Comprehensive Programming

As part of NaTCAP or otherwise, and for aid projects in general as well as for technical cooperation, many voices have begun to call for replacement of the project-by-project approach by programme approaches. Some commentators mean by this something close to general budget support by donors. Others, more numerous, understand it to involve the planning and budgeting of technical cooperation—the last phase of the NaTCAP process. Most seem to mean that technical cooperation and other inputs should be embedded in sectoral or subsectoral programmes, slices of which donors could then finance.

The growing support for the programme idea is suggested by the fact that the General Assembly of the United Nations has adopted resolutions calling for a shift away from project-by-project aid arrangements to a programme approach. The new consensus is reflected also in the most recent DAC statement on technical cooperation:

> Increased emphasis should be given in the planning, selection and design of technical cooperation activities to a programme rather than a project-by-project approach. The programme approach should be based on thematic, sector-wide, multi-disciplinary and often multi-donor actions.... [Aid agencies have agreed that they] ... will plan and manage ... aid increasingly in the context of coordinated support for larger sectoral programmes, objectives and policies. This principle should apply with particular force to Technical Cooperation. The effectiveness of technical cooperation has suffered from a piecemeal approach.[25]

Implicit in this analysis is the assumption that governments should and would begin by developing detailed sectoral and subsectoral programmes to achieve national priority goals. These would contain specific, usually quantified output targets (500 kilometers of class A road construction during the next three years, 1,000 kilometers of road rehabilitation, repair of 85 bridges, training of 25 civil engineers, privatization of maintenance workshops, and the like). The programmes would also indicate input requirements—so much equipment (new graders or bulldozers), materials (asphalt and gravel), and labor, including technical assistance. Local resources available for financing the programme would be estimated. Donors would be asked to fill the gaps.

There is no doubt that much of the criticism of the project approach that underlies these proposals for programming is valid.[26] The use of projects as the main vehicle for technical cooperation has had numerous unfortunate (and unintended) consequences. Concentration on the success of "their" projects is one factor that has led donors to assume increasing control over design and implementation processes as a way to circumvent or neutralize inefficient (sometimes corrupt) officials.

The widespread use of project implementation units is illustrative. These units—often situated within ministries—have considerable autonomy, narrow focus, and access to donor resources for staff and supplies. Because donors provide these resources, they can control the project units. In civil service environments where real salary levels have eroded, the project units are able to recruit good people and maintain adequate supporting budgets, unlike regular line units in their home agency.

Proliferation of such largely independent project implementation units significantly distorts work-force allocations and weakens already feeble technical ministries. It increases coordinating needs while reducing coordinating capacity by the host agency. It creates discontent by setting up dual salary structures. The synergism that should result from close coordination of related projects and activities is often lost. And when donor financing stops, so too does support for the autonomous project unit; their sustainability is everywhere a problem.

Reliance on individual projects, then, has had many negative effects, from exacerbated administrative disorder and reduced prospects for sustainability to the growth of micromanaged assistance programmes and increased donor influence in the aid process. But the project concept per se is not the culprit, and the problems evoked are not likely to disappear by adopting a programme approach. In fact, programmes have to be made operational before they can be implemented, and the vehicle for doing so remains the project.

The project, after all, continues to be the elemental unit of development action—an activity with a well-defined beginning and end, with specific, usually quantifiable outputs, and with measurable inputs or costs. Before money can be spent for development purposes and people hired or assigned, equipment purchased, and construction undertaken, there have to be concrete and detailed plans. There must be designs, terms of reference and job descriptions, course curricula, equipment specifications, and much more—all the humdrum, nitty-gritty stuff that is the raw material of effective implementation of any programme.

This is not to deny that there may be economies of scale in project design and management that can be captured by a programme approach. It is intuitively clear that in a rural water supply effort planning and management costs per well are likely to be cheaper under a programme encompassing 100 village wells than they are in the case of individual well projects covering many fewer villages. The same is probably true in cases where broader programming approaches have actually been introduced, such as a rural roads programme in Kenya and vaccination programmes throughout the world.

For these reasons and others, it is desirable to put projects into a programme context. Focusing on the programme permits better sectoral planning, and it may remedy some of the deficiencies of the present project-based system. In some cases at least, it can reduce the need for detailed project planning and preparation and can lighten project management burdens. But somewhere along the line project identification and preparation will be necessary: lots of nuts and bolts will still have to be put together. And there is another important consideration: in the least developed countries, diseconomies of scale set in quickly—for example, because communication and coordination are difficult—so turning projects into programmes by multiplying their components has its limits.

The underlying problem is here. Programme approaches will probably not significantly reduce demands on local project preparation and management capacities. And they make stronger demands on planning and coordinating capacities. To make them work, the recipient governments must have the capacity to specify sectoral priorities and place them in an acceptable macroeconomic framework; to identify, design, cost, and implement the projects and activities that make up the programme; and to integrate and coordinate their implementation. This capacity is embryonic in most of the region.

This fundamental dilemma affects not only the comprehensive programming approach but programming as it is envisaged in the NaTCAP process—that is, the preparation of technical cooperation programmes that are inventories of approved (priority) technical cooperation projects, phased and costed. In their most developed form, these are integrated into the planning-budgeting process—technical cooperation equivalents of the public investment programmes now in use for capital and related investment.

Despite extensive forward movement under NaTCAP, the transition from analysis and organizational strengthening to effective programming of technical cooperation has proved difficult. Five countries have moved to the programming phase: Burundi, Guinea, Guinea-Bissau, Mozambique, and Swaziland. Several others are entering it. In Guinea, the process has gone furthest; a near-mirror-image of the Three-Year Public Investment Programme has been prepared for technical cooperation.

Various factors explain slow progress. One is frequent lack of agreement about the purpose of programming technical cooperation, or a lack of urgency to control inputs that seem to have little opportunity cost. There is also a lot of uncertainty about how to do the programming in practice.

By far the most important factor, however, has been weak public sector organization and scarcity of skilled and motivated civil servants. The limited capacity to plan, design, supervise, and monitor technical cooperation projects makes the idea of comprehensive programming moot at this stage. Even more limited programming requires the screening, inventorying, and programming of expenditures for hundreds of technical cooperation projects—tasks that are beyond the institutional reach of most African public sectors.

Full-scale introduction of technical cooperation programming in the context of a NaTCAP exercise or otherwise may not be feasible. Given limited administrative capacities and multiple objectives, national authorities may have to make trade-offs among reform goals. Moreover, if it is poorly done, technical cooperation programming might even have unintended negative effects.

For example, one potential danger from incorporating technical cooperation programmes with so many constituent projects into the public investment programming process is that expenditure controls might be weakened. Technical cooperation projects consist largely of operating costs. If they are processed separately from the regular operating budget, financial management can be more complicated and expenditure control more difficult. This kind of effect often occurs when development budgets contain both capital and recurrent items. Spending ministries often seek to put in the development budget recurrent items that would be rejected if submitted through the regular budget process. This might happen in the course of a technical cooperation personnel reform if expenditure screening and monitoring is less effective in the technical cooperation programme budget than in the regular budget.

Similarly, a trade-off probably exists between comprehensive coverage of projects in a technical cooperation programme and improvement of project quality because the efficiency of project selection procedures is likely to be affected by the number of projects to be screened. So while inclusion of all projects is desirable for purposes of macroeconomic planning, this objective may have to be sacrificed in favor of a more partial approach, one that focuses on proper evaluation, financing, and implementation of projects that are above a certain size or otherwise of high priority.

Along the same lines, efficiency of technical cooperation use and of aid resources might be increased by joint reassessment of approved projects. This is true of capital projects as well as technical cooperation. Thus, a recent World Bank paper argues that, in most heavily aided countries, public investment programmes are overloaded with projects and need to be "unclogged." The authors urge the creation of small task forces, as was done in Ghana, that can work with individual donors to scale back or cancel low-priority projects. This would release resources (aid, local budget money, and implementing capacity) for higher-priority programmes.[27] The NaTCAP experience in Guinea points in the same direction: the programming exercise there contributed to cuts of 30 percent in technical assistance personnel expenditures and cuts of 25 percent in overall technical cooperation costs.[28]

TOWARD A GRADUALIST REFORM STRATEGY

This brief review of existing approaches to technical cooperation reform suggests some general observations on reform strategies as well as some specific proposals to increase the prospects for success.

The concern throughout this chapter has been with reform of technical cooperation management in its internal dimension; we have considered only in passing the external environment within which the technical cooperation system operates. There will be more about that later. With respect to internal reform of technical cooperation management, the conclusion emerges strongly that there are no practical alternatives to gradualism. This is evident from the experience thus far with the most extensive reform efforts—those associated with the NaTCAPs, which in practice have followed exploratory and gradual approaches. This conclusion also emerges from reflections on experience with institutional reform in general. The evidence suggests that quick reform of deep-rooted institutional systems is unlikely. In the present donor-led system of technical cooperation management, neither donors nor recipients are likely to accept or be able to implement basic changes in its internal structure. Most donors are unwilling to relinquish control over technical cooperation management at one fell swoop. And most recipients need time to develop the capacity to take on vast new management responsibilities.

A more realistic approach is to disaggregate—to identify those dimensions of the management problem that create the biggest obstacles to the achievement of development goals and that are, at the same time, the most amenable to change. Once specific constraints are identified, it is necessary to decide which ones deserve priority attention. To guide this decision, some rough and ready criteria can be applied:

- Which problems, once resolved, are likely to have the largest impact on the problem of technical cooperation management?
- What, if any, is a logical sequence in which to attack the identified problems? Is the resolution of one problem a prerequisite to the resolution of another?
- Which problems are the most amenable to change? In particular, on which problems is there likely to exist a solid consensus from all stakeholders—both on the donor and on the recipient side—about the need for change? For example, there is probably more consensus about, and therefore less resistance to, the idea of improving information systems than there is to the idea of untying technical cooperation resources.

These kinds of considerations suggest the following propositions regarding targets and sequencing of technical cooperation reform efforts.

- The generation of more and better information about technical cooperation is the highest-priority task, by all criteria. It has the greatest potential impact, is a prerequisite to many other changes, and is relatively easy to introduce. The kinds of information emphasized in NaTCAPs are indicative of what should be done: inventories of technical cooperation projects; estimates of total and sectoral technical cooperation costs, domestic and external; policy analyses; and diagnostic studies.
- As with capital projects, improvement of the quality of technical cooperation expenditures is a high-priority objective. It depends on introduction of numerous organizational and procedural changes, which should receive early attention in any reform programme: for example, creating a central unit responsible for technical cooperation coordination; creating or

reinforcing capacity to develop sectoral programmes with well-defined priorities and projects; and strengthening vetting procedures (creation of a single channel of approval, for example).

- Global assessments of needs (whether called "institutional," "capacity," or "manpower" assessments) can be useful. But there are many conceptual problems inherent in the attempt to define work-force needs; the debris of the manpower plans of the 1960s is rich in lessons.[29] Manpower and related types of assessments should therefore be done lightly. This should not be difficult, since the main lines of public sector skill scarcities are readily apparent in most countries. At a later stage in reform processes, assessments of greater refinement may be needed. In any event, in most cases manpower assessments do not meet the first two criteria mentioned above: their impact is uncertain and they are not prerequisites for many reforms. They can also be time consuming and expensive.[30]

- In countries with acute administrative weaknesses, technical cooperation programming need not aim at a programme that is a mirror image of the public investment programme. It would be desirable to start slowly, perhaps by developing a shadow (that is, unofficial) technical cooperation programme for the first year or two. A partial programme might also be tactically appropriate—in other words, starting out with a few key sectors. Ministries of education, health, and agriculture, and the core economic agencies, are the sites of most technical cooperation. Reformers might therefore begin by programming one or several of these, again perhaps on a shadow or informal basis.

- Efforts to strengthen local management at the micro level have high priority. This should mean first of all strengthening the capacity for managing technical cooperation in the user agency, where it is almost everywhere deficient. Improvements at this level are likely to have the most immediate development impact. Also, despite the limited scope of such micro-level

improvements initially, their demonstration effect could spread vertically as well as horizontally.
- It is probably more promising to focus on free-standing technical cooperation projects and to give only secondary attention to technical cooperation incorporated in capital (or investment) projects. The continued use of expatriate personnel in capital projects appears to stem, at least in part, from the need on the part of donors to control the projects' investment resources. This should be much less of a problem in free-standing projects, which often have no resources to control except the technical assistance itself.

The paradox that runs through this chapter has been widely noted: the deficiencies of technical cooperation management arise largely from donor dominance of the process at all stages, but recipient capacities are too limited to allow effective takeover from donors. The decisions about technical cooperation acquisition and use take place in a framework in which the rules of the donors prevail, from project identification to implementation and beyond. The donors know the rules best, have fitted their administrative structures to accord with their idiosyncracies, and adapted to them most effectively.

In these circumstances, it is hardly ever worthwhile for national officials to submit project proposals for donor financing. They are too likely to be rejected because they are not written in the right language, or packaged with the proper justifications. As the NaTCAP studies have shown, African officials often adopt the attitude that "beggars can't be choosers." One among the many aspects of the passive dependency that results is the feeling that there is no point in trying to develop national procedures and criteria for technical cooperation projects. And as African administrations have become weaker in recent years, their ability as well as their incentive to master this process becomes weaker. A vicious circle sets in and donors do more and more. They put advisers in planning ministries to help move the papers, and they hire nationals to work in their own resident missions to do the same.

Progress has been and can be made in transferring management authority to local hands. But as long as reformers have to take the environment surrounding the technical cooperation system as given, their task is immensely difficult. The most important elements of the environment are the existence of high levels of external assistance, multiple donors, weak public sector management capacities, and a market in which prices and opportunity costs play very little role in shaping decisions. In effect, a weak management system is being asked to allocate technical cooperation purely by administrative means in a situation in which market discipline does not exist or market forces operate perversely.

Any reform strategy therefore has to include efforts to change the environment within which the technical cooperation system functions. This means creating a more effective market for technical cooperation, most importantly, by introducing prices into the system. It also means coming to grips with the institutional constraints and in particular with the deteriorating performance of African civil services caused largely by inadequate incentive systems. These external changes are addressed in the next two chapters.

NOTES

1. OECD, Development Assistance Committee, *Development Cooperation, 1990*, Paris, 1990, Table 37.

2. World Bank, *World Development Report 1991*, Tables 1, 11, 20.

3. For a good summary of aid management and coordination issues see S. Lister and M. Stevens, "Aid Coordination and Management," World Bank, 1992.

4. The term "unequal partnership" was put into common use in the course of a study of technical cooperation in East and Southern Africa financed by the Nordic aid agencies. See Kjell J. Havnevik, "Unequal Partners: The Role of Donors and Recipients in the Identification, Design and Implementation of Technical Cooperation Projects," UNDP internal paper, May 1990.

5. The terms "spending agency," "implementing agency," and "technical ministry" are used interchangeably.

6. These included 42 bilateral donors, 25 multilateral donors, and 15 nongovernmental organizations. It is safe to say that the number is much larger now if for no other reason than that the number of nongovernmental organizations has vastly increased in the last 10 years. Elliot Morss, "Institutional Destruction Resulting from Donor and Project Proliferation in Sub-Saharan African Countries," *World Development*, May/June 1984.

7. Management in this sense includes management of the entire program and project cycle: identification, programming, design, implementation, and evaluation.

8. Governments may not, for example, be keen on accepting advice from "resident advisers" but may find them useful in carrying out politically sensitive studies, in acting as liaisons with funding agencies, or simply as extra hired help to cope with the workload.

9. One of the busiest times for design teams engaged by A.I.D. is the spring and summer as the pressure mounts to prepare projects that can be approved in Washington and for which funds can be committed before the end of the U.S. fiscal year, which is September 30.

10. The Maastricht report describes one of its case study findings as follows. "The Tanzania case study ascribes the poor record of reforming TC management mainly to 'the lack of a unanimously sustained political and administrative (managerial) commitment to change ways and processes of TC.' Important groups and sections 'still do not see the need for change,' hiding away in a 'wait-and-see attitude.' . . . The Burundi case study also refers to the existence of 'un esprit attentiste d'assiste (a passivity due to a psychology of dependency).'" J. Bossuyt, G. Laporte, and F. van Hoek, "New Avenues for Technical Cooperation in Africa," European Centre for Development Policy Management, Occasional Paper, Maastricht, 1992, p. 56.

11. See Beatrice Buyck, "The Bank's Use of Technical Assistance for Institutional Development," World Bank, PRE Working Paper, WPS 578, January 1991; and Samuel Paul, "Institutional Analysis in World Bank Projects," paper for World Bank Conference on Institutional Development and the World Bank, December 1989.

12. See the definitions and discussion of institutional development in Chapter Two. See also the World Bank report, "Managing Technical Assistance in the 1990's," 1991, which defines "institutions" as the

"norms of behavior that govern transactions, or the way people deal with each other: property rights, contracts, regulations."

13. After noting the desirability of "government execution" of UNDP-financed technical cooperation projects, a recent policy statement observes: "Governments are not always inclined to assume responsibility for the execution of UNDP assistance to their projects. . . . In one case a suggestion for Government execution of a project was turned down by the Government Department concerned on the grounds that the Department did not have the capability of administering the project nor facilities to recruit and supervise expatriate staff. . . . One Resident Representative reported that: 'the only fundamental new dimension to which the Government has objected is Government execution of projects in the near future . . . (because) it would draw too heavily on the scarce high-level national ranks.'" United Nations Development Programme, "Programme and Projects Manual, Overall Framework of UNDP Mandate," February 1988, par. 19.

14. Lister and Stevens, 1992.

15. See Martin E. Adams, "Aid Coordination in Africa: A Review," in *Development Policy Review*, vol. 7, 1989, p. 186.

16. This list comes mainly from Lister and Stevens, 1992, pp. 22-24.

17. R. Cassen and Associates, *Does Aid Work?* 1986, p. 208. Cited in J. Bossuyt et al., 1992, p. 53.

18. Bossuyt et al., 1992, p. 53. According to one writer, in Sudan "bilateral donors tend to be law unto themselves, preferring to maintain their freedom to disburse funds in the most opportune manner." Adams, 1989, p. 188.

19. The reasons for this are explained in Chapter Two.

20. See Bossuyt et al., 1992, pp. 26-27.

21. Background is provided in Carlos Lopes, "L'expérience des NaTCAPS," internal UNDP paper, 1990; and in an interview by P.C. Damiba in *African Recovery*, December 1989. See also, UNDP, Regional Bureau for Africa, "NaTCAP Methodology," May 1989.

22. The Maastricht report states: "Compared to other coordination efforts, NaTCAP has achieved a high degree of internalization, involving a wide array of national actors and stakeholders. Precedence is clearly given to long-term institutionalization over quick results (e.g., by bringing in a large number of expatriates to write reports and manage the process)." Bossuyt et al., 1992, p. 55.

23. Maurice Williams et al., "Evaluation of National Technical Cooperation Assessment and Programmes (NaTCAP)" for the United Nations Development Programme, Regional Bureau for Africa, vol. I., February 1991; Bossuyt et al., 1992, pp. 53-57; World Bank, "Report of the Technical Assistance Review Task Force," December 1991, par. 3.20.

24. In Malawi, the Department of Personnel Management and Training was the target of institutional support aimed at making it such a focal point. But donors see the department as ineffective and beyond easy reinforcement; they doubt its capacity to serve as a focal point for technical cooperation management. M. Williams and A. Nikol, "Final Report, NaTCAP," UNDP Evaluation Mission, Malawi, December 12, 1990.

25. OECD, DAC, "Principles for New Orientations for Technical Cooperation," Paris, December 1991, pars. 27-28.

26. This section draws on a background paper prepared for this study by Bengt Sandberg, "A Programme Approach," 1991; and on Bruce Jenks, "Toward a Programme Approach," UNDP internal paper, 1989.

27. Lister and Stevens, 1992, p. 20.

28. Williams et al., 1991, vol. II., p. 11; and interviews with UNDP staff in Conakry, May 1992.

29. Robinson G. Hollister, "A Perspective on the Role of Manpower Analysis and Planning in Developing Countries," World Bank Staff Working Paper No. 624, 1987.

30. In Malawi, under a World Bank technical assistance project, a human resources inventory was undertaken in the late 1980s. Although some benefits resulted, the operational usefulness of the study was minimal.

5

CREATING A MORE EFFECTIVE MARKET FOR TECHNICAL COOPERATION

Most development practitioners do not think in terms of a market for technical cooperation. They rarely consider technical cooperation a service that has definable supply and demand functions and a price. Nonetheless, like any scarce commodity, technical cooperation has costs that somebody has to pay. Analysis of the determinants of technical cooperation supply and demand, the way technical cooperation markets function, and the role of costs and prices in this form of aid is essential for an understanding of the general problem of technical cooperation efficiency and effectiveness.

The technical cooperation market is highly imperfect; it has several special features that help explain the dysfunctional bureaucratic behavior and poor results that are now widely deplored. First and most important, costs and prices play only a small role in the determination of supply and demand, encouraging economically irrational use of technical cooperation resources. Opportunity cost consciousness is stunted on both supply and demand sides.

Second, buyers (takers) of technical cooperation have limited flexibility in choosing the mix of inputs they want; technical cooperation packages are bundled by donors (suppliers) and cannot be unbundled by recipients (demanders). Moreover, regulations of some donors restrict the right of recipients to choose between local and imported skills, even when local skills are available, would be cheaper to use, and can be employed suitably in donor-financed projects.

These distortions or failures in the market for technical cooperation are of considerable significance in shaping the behavior of all the parties to the technical cooperation transaction. Yet until recently, most have received little attention by analysts or practitioners.

This is now beginning to change. Practitioners, more and more frequently, express uneasiness about the "free good" character of technical cooperation. This malaise is reflected in the December 1991 DAC report on new orientations for technical cooperation, which urges that greater attention be paid to costs and cost-effectiveness by both donors and recipients.[1] But cost and related issues remain too much in the shadows. And the nature of the market for technical cooperation has not been the subject of much systematic analysis. Thus, a background report prepared for a recent international conference hardly mentions it at all.[2] The DAC guideline does address it, but not in terms of market imperfections and market-oriented solutions.

The 1991 report of the World Bank's Technical Assistance Review Task Force has a useful annex that mentions market imperfections.[3] But its focus is on the supply-driven aspect of technical cooperation and it recommends only administrative measures as solutions. Moreover, it comments favorably on the use of grants to finance technical cooperation, and itself proposes a new World Bank grant facility for financing institutional-development activities, mostly technical cooperation. These changes will not reduce imperfections in the technical cooperation market; rather, they risk enhancing them.

In this chapter we consider first the main failures or imperfections in the technical cooperation market: lack of cost consciousness among macroeconomic policy makers and operating agency users of technical cooperation; the divergent costs and benefits between users and payers; weak price and cost consciousness on the supply side; and the limited ability of buyers to untie the aid package made up of technical assistance personnel, supplies and equipment, and training. We then

CHAPTER 5 Creating a More Effective Market for Technical Cooperation 167

propose ways to make these markets more responsive to economic concerns about resource use and more flexible from the point of view of technical cooperation users.

IMPERFECTIONS IN THE MARKET

The provision of technical assistance takes place in a market in which costs and prices play little or no role in influencing demand and supply and in allocating the service (technical cooperation) between user agencies. Several distinctions are helpful in analyzing how this market works.

- Market actors fall into two groups, recipients and donors, with each in turn divided into decision makers in the core economic agencies (ministries of finance and planning and central banks on the one side, aid agency headquarters on the other) and the decentralized implementing bodies—recipient operating agencies (technical ministries and parastatals) and donor aid field missions or project units.
- The macroeconomic problem and macro-decision making have to be distinguished from the micro problem. The former concerns the overall volume of technical cooperation and its sectoral allocation; the latter, the technical cooperation projects.
- Money or financial costs should be distinguished from real, social, or opportunity costs. The former are the outlays incurred in the production of any good or service. The latter are what the host society gives up as a result of producing (or buying) that good or service.[4]

Lack of Cost Consciousness at the Macro Level

The distinctions outlined above clear the way to an understanding of how and why technical cooperation markets function poorly. First, at the macroeconomic level, decision makers in the core economic agencies of the recipient country are rarely preoccupied with the overall volume of technical cooperation. This is so despite the heavy weight of technical cooperation in the aggregate—about a quarter of total official development assistance to Sub-Saharan Africa in recent years, and more to many individual countries.⁵ The main reason for this lack of preoccupation is that neither money nor real costs to the local economy are perceived to be significant. One consequence is that the incentive to economize on the volume of technical cooperation is minimal. For finance or planning ministries to reject assistance in the form of technical cooperation projects seems unwise: this rejection brings small financial gains while economic and political bureaucratic costs might be large.

Are finance or planning ministers right not to worry about the aggregate amount of technical cooperation they seek or accept? Yes and no. Some financial cost is paid by many recipient governments and borne by the domestic economy. Not all technical cooperation is given as grants; the World Bank and many non-OECD donors offer it on reimbursable terms. Even for donors offering technical cooperation on grant terms, many (the French, Swiss, Belgians, and Portuguese, for example) ask the recipient governments to make compensatory aggregate payments for technical assistance personnel or in-kind contributions, notably housing. In parts of Africa where donors provided topping-off salaries for technical assistance personnel (some countries in Southern and East Africa), host governments had to pay base salary costs, but these arrangements are no longer common.

The bulk of technical cooperation is provided on grant terms, with few associated costs. Even in the past, the money costs paid by the host country were in most cases modest, compared with the market prices of the services provided. Moreover, they were often unpaid when budgets came under stress. During the 1980s, sympathetic donors tended to relieve hard-pressed governments of these obligations.⁶

The real issues relate to opportunity costs. These are of two kinds. The first consists of transaction costs—defining terms of reference, negotiating the aid agreements, and managing the technical assistance personnel once in place. These are not burdensome, and in any case are mainly borne not by macro policy makers but by the operating agencies that house the technical assistance personnel. The major opportunity cost is in sacrificed official development assistance. There is surely some trade-off between technical cooperation now and other official development assistance now, or between technical cooperation now and technical cooperation or other assistance later. The question is how substantial is it likely to be.

The answer depends on the way donors fix country programme size and budget aid resources among programmes within countries. If donor aid organizations set (or are given) a specified amount for each country programme, without predetermined proportions for policy lending, capital or development assistance, and technical cooperation, then country allocations are flexible and each component is fungible. In this case, the opportunity costs of technical cooperation are likely to be substantial, even in the short run.

The trouble is that little is known about this question of how aid agencies do their intercountry allocations and their internal allocations among different types of assistance. Under one set of assumptions (not implausible for some donors), technical cooperation opportunity costs would be small—for example, if the donor's global spending targets for a given country is not met, or if the donor's country spending targets are flexible, so that more technical cooperation does not force a cutback in other types of assistance. Under another set of assumptions, it is likely that most aid agencies can move resources between different types of aid (from technical cooperation to policy lending, for example), especially over the medium term. In this case, fungibility and hence technical cooperation opportunity costs are greater.[7]

Whatever the degree of fungibility of aid resources, other considerations lead to an expectation of small technical cooperation opportunity costs. One is that much technical cooperation is complementary to other assistance, not competitive with it. This is true of all technical cooperation related to capital projects—the civil engineers who design and oversee construction of roads, the water control specialists in irrigation projects, and so forth. It is also true of policy-linked technical cooperation—all the free-standing technical cooperation projects tied to structural adjustment loans, for example, or the financial specialists who may be included in an energy sector reform project. For these loans, reductions in technical cooperation might entail reductions in other kinds of assistance, not increases.

More important, in the longer term, the level and content of future aid flows are likely to depend only partly, and to a highly uncertain extent, on how much technical cooperation and other official development assistance they receive today. Many other factors will determine the aid received: the way the world changes, the nature of shifts in donor policy, the country's own policy performance and political evolution, the ability of the country to retain old political friends and make new ones, and its capacity to generate good projects. Decision makers could not be faulted if they hesitated to cut technical cooperation today in the expectation that there would be more aid (technical cooperation or other) in the future.

This, then, is one important market distortion: low financial costs and uncertain, perhaps low, opportunity costs take away from macroeconomic policy makers the economic incentive to restrict aggregate inflows of technical cooperation.

Divergent Costs and Benefits Between Users and Payers

A second key aspect of market malfunction is the divorce between actors who benefit from technical cooperation and those who pay most of the costs incurred by the presence of expatriate technical assistance personnel. As noted earlier, technical cooperation decision making on the demand side is decentralized, and individual bureaus or ministries are mainly responsible for projects. For these decision agencies, technical cooperation is free or nearly free in terms of budget costs. In real terms, the user agency may incur some search costs to line up financing and some nuisance or inconvenience costs if things go badly, but these are minor in magnitude and easily dealt with.[8]

In return, the user agency is the main direct beneficiary of the technical cooperation presence, and its benefits are numerous. The user agency receives skills not otherwise obtainable because of scarcity or budget constraints. Technical cooperation brings vital materials and supplies—vehicles, photocopying machines, computers, paper, sometimes salary supplements, a window to the world, and a channel to donors. Managers of user agencies sometimes find it easier to work with expatriate personnel because they are outside the arena of domestic politics and bureaucratic jockeying. Expatriates often identify with local agency management and are responsive to its needs for technical and other support.

The second demand-side market distortion, then, is the gap between costs and benefits at the project level, benefits swamping whatever financial and real costs exist. The actors who benefit (operating agencies) are not the actors who pay these costs, in any case. This helps explain the existence of a lively demand for technical cooperation. The demand is effective because African governments have been unable to impose economic rationality on the market by administrative means—by planning and programming technical cooperation for priority needs, and by imposing vetting procedures and machinery that can make the decision process more orderly.[9]

Weak Cost-Price Signals on the Supply Side

Cost plays some role on the supply side. In requests for proposals (when technical cooperation projects are put up for bids), cost is one criterion for selecting winning contractors. But given a country programme, the offer of technical cooperation projects is not responsive to project costs—in other words, the supply of each donor's projects is price-inelastic. This is so for several reasons.

- Donors seek to satisfy many different objectives with technical cooperation, and these often make cost considerations secondary. As on the demand side, much technical cooperation is complementary to investment projects and policy loans. Its cost is often a small proportion of total project cost, and its benefits in expediting project implementation are highly valued. So donor agencies at all levels—national headquarters, local aid missions, and project implementing units—regard this kind of technical cooperation as decidedly cost-effective.
- In both national headquarters and resident local missions, technical cooperation is commonly seen as an activity with multiple positive externalities (spillover benefits).
- Donors often see expatriate technical assistance personnel as performing many functions in addition to those of adviser or

technical expert. These personnel can act as controllers, ensuring better project financial management; as expediters, speeding implementation of projects; and as facilitators, providing access to officials and knowledge about the way things work. Most of these ancillary functions grow in importance as public sector capacity declines.

- Technical cooperation projects can provide benefits to and assist private firms, universities, research and policy institutes, nongovernmental organizations, and other agencies and individuals from the donor country.
- Institutional factors have great influence in shaping the decision to use technical cooperation.
  — Strong supply-side pressure springs from deep-rooted habits of project design—notably, the tendency in donor agencies and among the consultants they hire to design complex projects and then find that technical assistance personnel are essential for their implementation.
  — In addition, many donors tend to recommend technical assistance personnel whenever institutional weaknesses are identified as important obstacles.[10] Because the "need" is thus said to be rooted in such fundamental structural constraints, cost considerations fall by the wayside.
  — A final contributing factor is the tendency among consulting firms that design projects to propose abundant expatriate personnel. This stems from habit, from the belief that it is essential for project success, and from the fact that it is profitable for the implementing consultants.[11]
- Donors often behave as though they are in a competitive market for projects, technical cooperation and other. And, at least in heavily aided countries, they are not wrong: the situation is often one of projects chasing recipient agencies. A technical cooperation project that leaves one donor indifferent will enthuse another. Suppliers, therefore, have incentives to compete by making projects more rather than less generous.

In such a market, in which the main actors seem to behave as though the use of technical cooperation involves giving nothing up or as though it has no opportunity cost, what is remarkable is not that there are so many technical assistance personnel in Sub-Saharan Africa, but that there are so few. How can this be explained?

On the recipient side, a number of factors appear to be at work that stay the hands of many of those who would seek or accept additional technical assistance personnel. In most countries of Sub-Saharan Africa, a political-psychological constraint exists—the unwillingness to have too many foreigners in key positions. National civil servants resist partly for this reason and partly because they are concerned about their own status. The tied character of much supply for technical cooperation (its link to capital projects or policy loans) reduces the buoyancy of this supply. Also, needs have declined in some key sectors, notably education, that formerly used technical assistance personnel intensively. Local people perceive that competent expatriate personnel are hard to find. And there is spreading recognition that the national economy pays real costs for technical cooperation.

On the donor side, similarly, constraining factors enter the equation. Stagnation or slow real growth of overall official development assistance may be a factor. More generally, it is not easy to conceive and design free-standing technical cooperation projects—that is, projects unlinked to capital investment projects or to policy loans and grants. These free-standing projects also are difficult to implement. Some donors are unwilling to finance these kinds of projects because their record of effectiveness is so poor. A related factor is increased awareness of the fungibility of aid resources, especially among staff at the headquarters of donor agencies, which leads to rejection of some low-priority technical cooperation proposals.

These constraining factors notwithstanding, the central point about technical cooperation incentive structures remains: the weakness of opportunity cost consciousness and a resulting tendency for there to be too much (that is, low-priority) technical cooperation; and the absence of a system of prices that would help allocate technical cooperation among sectors and programmes.

The market for technical cooperation, then, has this curious aspect: most actors in the market seem to behave as though the use of technical cooperation involves neither financial cost nor real cost. In reality, technical cooperation does have the financial costs noted above, despite the fact that most of it is provided on grant terms and many donors no longer request local cost contributions. And local economies do bear some opportunity costs, even though these are of uncertain magnitude.

Whether opportunity costs are small or large from the perspective of the national economy, for the user agency the financial and real costs of technical cooperation are very low or nonexistent. The user agency suffers no budgetary deduction when it takes on additional technical cooperation, and the real costs it incurs in managing the technical assistance personnel are small. The user agency nonetheless garners many benefits from this aid.

It is not surprising, then, that most recipient governments display little urgency to economize in the use of grant-financed technical cooperation. Nor are donor agencies conscious about the costs of technical cooperation. They are inclined to push technical cooperation for its direct and indirect benefits. Recipients tend to find these donor initiatives acceptable, and put forward other proposals on their own. The prevailing incentives contribute to proliferation of projects, minimal local ownership, haphazard linking of technical cooperation with national priorities, and wasteful retention of unsatisfactory or little-used technical assistance personnel.

Limited Flexibility for Buyers

Freedom of choice for the recipient country is regularly and systematically restricted in the technical cooperation market. Much technical cooperation, first of all, comes already packaged in investment projects. It can be challenged and disentangled but with difficulty. Some free-standing technical cooperation, similarly, is part of policy lending; the World Bank, IMF, and other donors sometimes insist on specific institutional strengthening as a loan condition.

Second, the three main elements of technical cooperation—personnel, equipment, and training—are also part of a package. They are normally tied—in other words, not available separately. If a bureau needs cars and copiers and training opportunities for its staff, it usually has to take technical assistance personnel as well.

Finally, in some instances the fungibility of technical cooperation resources is restricted by donor rules (see Box 5.1). Many bilaterals and the European Community, for example, strongly favor consultants from the donor country. It is true that donor country consulting firms are increasingly obliged to team with recipient country consultants. But the local consultants are almost always a small minority. When recruitment is not through consulting firms, recipient governments are normally not free to choose in their use of technical cooperation resources between local experts and expatriates.

These restrictions on hiring local staff are increasingly vexing to recipient country authorities. Forced by budget austerity (reflected in IMF or World Bank agreements) to forgo hiring of nationals, government agencies take on externally financed personnel while new university graduates and other trained local people are without jobs.

There is little question that, in a more competitive market for technical cooperation, one in which host country preferences were given greater weight, the result in most African countries would be less technical cooperation and more capital (or similar development) assistance. The mix of technical cooperation project components would be significantly different than it is now. Fewer personnel would be used, and more would be spent on supplies, equipment, and training. And of the personnel hired, more would be nationals and fewer would be resident expatriates. This at least can be inferred from the NaTCAP studies mentioned earlier and from such surveys as those done by the World Bank's recent Task Force on Technical Assistance.[12]

INTRODUCING MARKET ELEMENTS

Build on Technical Cooperation Programmes

Without widespread awareness of opportunity costs by donors and decision makers in host countries, more rational use of technical cooperation resources will be extremely difficult to bring about. As we

BOX 5.1

OBSTACLES TO HIRING LOCAL PROFESSIONALS

A.I.D. has a formal policy of encouraging the use of local professionals in A.I.D.-financed projects. For example, a recent long-term contract with a U.S. consulting firm—for provision of evaluation services—has the following clauses: "To the extent possible, and especially given the requirement for field-based data collection efforts, the contractor shall utilize cooperating country personnel." The contract specifies that out of the total 420.5 person-months that the consulting firm is to provide, 120 person-months should be provided by national experts.

This is certainly an encouragement to local hiring of professionals. But its impact is diluted by contradictory clauses in the contract, by A.I.D.'s contract management procedures, and by consulting firm practices.

A later clause in the contract discourages local hires. It states: "The Contractor will ensure, to the extent feasible, that there is a continuity of technical staff throughout the life of a major assessment." Because the assessments in question involve visits to numerous countries in different parts of the world, nationals hired to do one-country assessments could not provide the continuity demanded.

A.I.D.'s contract management procedures require that the salaries of personnel, including foreign nationals, be approved in advance of their use on any assignment by the A.I.D. contracting officer. Approval is obtained by submitting an Employee Biographical Data form (EBD), containing the salary history of the proposed individual and adorned with his or her original signature, to the A.I.D. contracting officer.

This requirement poses several disincentives to the use of foreign nationals. First, local professionals frequently do not have salary histories that can justify expected fees. A.I.D. contracting officers often reject salaries proposed for consultants in such cases. In the rush to meet deadlines for proposal submission, U.S. consulting firms may assume the risk of negotiating an employment agreement with a foreign national at a particular daily rate, only to learn later that the A.I.D. contracting officer has approved the use of the consultant at a lower rate. The consulting firm then has to make up the difference.

Second, distance makes it more difficult to obtain an EBD with the required original signature, at least in time to meet deadlines.

Third, many foreign national consultants work for local consulting firms, which means that obtaining their services requires a subcontract with the local firm. Subcontracts have to be negotiated and then approved by A.I.D., a process that takes several months at least, which usually makes it infeasible given the tight recruitment deadlines that are common.

BOX 5.1 — Continued

> With respect to consulting firm practices, a for-profit consulting firm has an incentive to use higher-paid consultants, which typically means expatriate rather than local personnel. This is so because overhead charges and fees are calculated, by A.I.D. rules and probably those of other aid agencies, as a percentage of the consultant's salary. Also, recipient country recruitment requires higher transactions costs—for example, more expensive and uncertain communications and possible travel.
>
> Recruitment of donor country consultants, finally, is much easier in all respects. Consulting firms work hard to build up rosters of trained and experienced consultants, most of them nationals of the donor country. These people are known to the firm, are easily reached, and are signed up relatively quickly. Recruiting recipient country consultants is obviously much more difficult.

saw in Chapter Four, changes are under way in many African countries that are strengthening technical cooperation management, and some of these (better data on technical cooperation, for example, and the introduction of technical cooperation programming) are establishing the foundations for the spread of cost-consciousness. These can be built upon to inject cost considerations more explicitly into the management process.

In the framework of the NaTCAP process, many African countries have introduced systematic programming of technical cooperation, represented by the preparation of a technical cooperation programme. Technical cooperation programming has been operational in Guinea since 1990, and Mozambique, Guinea-Bissau, and Burundi since 1991. It began earlier in Swaziland, predating the NaTCAP process. The technical cooperation programme involves, among other things, the identification of ongoing and new technical cooperation projects, and estimates of how much they cost. The programmes thus put building blocks in place on the technical cooperation side that mirror those that exist for investment programming—in particular, the rolling public investment programmes that have spread throughout Africa in recent years.

Public investment programmes usually show total project costs, foreign financed and local, and give budgeted expenditures for a three-year period. The first year of a public investment programme typically includes only projects that are likely to begin implementation in the current budget year. This part of the programme becomes integrated into the national budget; it is often called the consolidated investment budget, "consolidated" because it is intended to be all-inclusive, incorporating foreign aid as well as domestic investment expenditures and investments of the parastatal sector.

The technical cooperation programmes move in the same direction. By establishing inventories and costing out existing technical cooperation projects, they make possible the incorporation of technical cooperation expenditures into the public investment programmes and into comprehensive development budgets. Thus the Swaziland Technical Cooperation Programme forms part of a consolidated development programme; the Guinea Technical Cooperation Programme will be similarly organized in 1993, and a comparable evolution is taking place elsewhere.

Greater cost awareness and more rational technical cooperation decision making are brought about by the introduction of technical cooperation programmes. The simple setting out of the full array of technical cooperation projects, with their costs and sources of financing, is an eye-opener for all concerned. It induces the authorities responsible for central coordination of technical cooperation to see it as a scarce resource and, in the process of putting it together, may force user agencies to justify the technical cooperation they have or want. The technical cooperation commission in Côte d'Ivoire for French technical assistance personnel seems to be having these effects.

In Guinea, responsible authorities claim that the existence of the technical cooperation programme has put a stop to the rapid proliferation of technical assistance personnel that took place after the change of regime in 1984 and has reduced duplication. The central technical cooperation coordinating agency in that country has begun to exercise genuine management control over this aid; it knows what is coming from where and what users are asking for. In Swaziland, similarly, the technical cooperation programme process made clearer to responsible officials that grant-financed technical cooperation resources are not free goods, and facilitated the budgeting of domestic counterpart funds for projects.

These are encouraging examples. They point the way to the future. But the technical cooperation programme process has to be improved. The inventories of projects that have been drawn up in many African countries have few of these positive effects. Part of the reason is that they are backward looking: they present past disbursements and not current programmes and future commitments. And, partly, their impact has been weak because they are often badly prepared, badly presented, and poorly disseminated. So the programmes have to be done more carefully, and they have to become more forward looking —more like the public investment programmes.

Introduce Prices

Better technical cooperation programmes should lead to stronger management of technical cooperation resources. But prices still lack. And without prices and more cost awareness, the task of the central coordinating authorities will remain extremely difficult. When foreign resources are grant financed and when domestic counterpart contributions are small or zero, propensities to proliferation will persist, the quality of technical cooperation projects will be uncertain, and adherence to sectoral priorities will be haphazard.

Of course, to the extent that the coordinating agency is able to evaluate technical cooperation proposals effectively, these problems are diminished. As we saw in Chapter Four, however, in many countries the project selection process remains weak, especially the vetting for project quality that is done by planning agencies. Even when the public investment programme is well established, investment project evaluation often remains mechanical. Thinly staffed ministries of planning and overburdened budget agencies are able to give only perfunctory screening to many projects and budget requests. Sectoral strategies are often poorly (or too generally) defined, making it difficult to set project priorities within sectors. Frequently, governments establish so many priorities, or define them so broadly, that almost anything can pass. Criteria for selection of institutional-development projects and those in the social sectors remain vague. All of this suggests that for the many countries to which this description applies, it will be some time before technical cooperation projects can be effectively vetted and adherence to national priorities assured.

Hence the importance of attaching a price to technical assistance personnel—making user agencies pay for it—and imposing a ceiling on technical cooperation allocations, thus creating the necessity for choice. One way to do this is to extend the concept of the technical cooperation programme by making the costed inventory of existing technical cooperation projects the basis of a national technical assistance budget, within which user agencies would have an allocation based on their programme submissions. The technical assistance budget would be the technical cooperation counterpart of the public investment programme, into which the budget would be integrated. It would show all technical cooperation projects, with their sources of financing. Its domestic financing component would be shown also in the regular state budget, in the section that shows allocation of new resources or the investment or development budget.

The technical assistance budget would be financed from annual allocations of uncommitted revenues. In the budget systems of most developing countries, allocation of these "new" resources is a highly strategic exercise. Most often, ministries receive their last year's operating budget, plus some part of their investment budget requests and whatever they can obtain from the pool of uncommitted resources that comes up for distribution each year. So if the allocation to technical cooperation were to take place in the framework of the bargaining over allocation of total uncommitted resources, it would enhance opportunity cost awareness and increase the rationality of technical cooperation decision making.

The total size of the annual allocation for technical assistance personnel, and its sectoral composition, would be determined in the same way as investment budgets are now determined. The aggregate needs for technical assistance personnel would be weighed against competing demands for the investment programme, for ongoing operations and maintenance, and for the operation of completed projects.

The key innovation would be to impose on user agencies a budget charge—for example, a percentage of the total person-month cost of personnel proposed. This user fee would be the equivalent of the so-called local counterpart contributions that many donors require in capital projects: a percentage of the total project cost that will be paid for in local currency by the recipient government.

One problem immediately arises: where would the money come from to pay the "fees" for technical assistance personnel that spending agencies now use without charge. The idea of making poor country governments pay for personnel goes against the grain, among recipients especially. The technical assistance budget nonetheless would have to be financed at least in part from domestic resources if the notion of choice is to be made credible. The budget could be financed partly from global allocations by donors. Although this might threaten the credibility of the exercise because donors would be financing local technical cooperation obligations through the back door, the process would still have some meaning. The attachment of a fee from domestic resources, coupled with the rule that all allocations of budgeted domestic resources must be approved by the budget and plan authorities, would encourage user agencies and their donor partners to submit all proposed technical cooperation projects to planning and budget agencies for formal vetting. User agencies (and their donor partners) would also have to be prepared to justify their projects to the budgeters, which is sure to have healthy consequences.

The technical cooperation programming exercise should yield a set of evaluated and approved technical cooperation projects for the first year, and a collection of project ideas or project outlines for the second and third years. The first-year programme would specify what projects are to be implemented, how much they would cost, and how they will be implemented—by long-term expatriate advisers, short-term foreign or local consultants, locals hired as contractual employees of the ministry, or other means. These decisions would have been made by the user agency, in dialogue with donors at earlier stages, when the project was in the second or third year of the three-year technical cooperation programme.

Unbundle the Technical Cooperation Package

Recipients should be able to take one or another element in the technical cooperation package without accepting expatriate personnel. And donors, with governments, should make a direct and comprehensive attack on the problem of support services and supplies. The scattered training inputs of projects should also receive more direct and systematic attention, and be integrated into national training strategies. Recipients should have greater freedom to use national staff for project design and implementation.

Studies done under NaTCAP auspices confirm the widespread observation that some African governments ask for or passively accept technical assistance personnel because they really want the equipment, supplies, and training opportunities that are tied to the personnel. This situation contributes to lack of local commitment to the personnel components, especially when expatriate technical assistance is at issue.

This market distortion or imposed constraint on resource use is best addressed by unbundling the technical cooperation package, separating personnel from at least some of its attachments. This is not only easy to do, but would also be highly desirable for reasons other than greater flexibility in using technical cooperation resources.

To some extent—in most countries very slight—technical cooperation projects and the related equipment and operating money it brings in have softened the impact of inadequate budget allocations and poor organization of support services—the lack of supplies, often grossly inadequate computer facilities and photocopying capacity, crippled transport capacity, and inability to repair and maintain government facilities and equipment. In many offices around the continent, the computers and computer paper of the technical cooperation personnel, their photocopying machines, and their access to vehicles allow operations to limp along. In their absence, governments would be forced to confront these basic problems directly.

Instead of continuing along this path, it is time to begin the search for long-term solutions. African governments, assisted by their external collaborators, should mount direct and comprehensive programmes to improve provision of these support services. It is not reasonable to continue to accept, in so many countries, the extraordinary spectacle of governments trying to operate in the presence of scarcities of the most elemental work tools of modern administration.

Much more money and imagination should go into free-standing technical cooperation and institution-building programmes aimed at problems of support services: for example, devising practical motor pool arrangements, secretarial pool systems, and arrangements for maintenance of computers and copying machines. Privatization of these services may present new possibilities for effective action; these are natural areas for contracting-out to private agents. In any event, these comprehensive programmes should replace the existing piecemeal, institutionally enfeebling arrangements whereby support services come attached to technical assistance personnel.

Institutional development of this kind is a difficult, long-term undertaking. As is pointed out in the next chapter, risks of failure are high, in this as in other areas of administrative reform. But the effort is essential. And in any event, whether or not the support services problem is dealt with comprehensively, technical cooperation projects need not be so tightly packaged. Governments should find it possible to request equipment or training without technical assistance personnel. Conversely, technical assistance personnel should be provided without its attachments. Long-term advisers and experts should be provided with what they need to do their jobs—personal vehicles, personal computers, small copying machines, or budgets for copying—but not with equipment to serve wider clienteles. The wider problems should be attacked frontally.

Employ a Mixed Strategy for Training

Training can be handled in various ways. In-service training is so much a part of the technical assistance presence that at least some of it should remain attached to the provision of technical assistance personnel. But it has to be done better. It is well known that training components of technical cooperation projects are frequently treated as of secondary importance—as an afterthought. Training is often hurriedly designed and poorly implemented. At the same time, when public agencies have developed integrated career development programmes and coherent in-service training schemes, training can be autonomously planned and implemented. Financing can be provided to the training office of the agency using the technical assistance personnel.

Fellowships for longer-term training could also go either way. All fellowship money could be channeled through a human resources development unit (if one exists) or a central scholarship fund that is established on a public sector or economy-wide basis. But the award of fellowships also can be a major incentive for public employees; their attachment to an institution-building technical cooperation project can induce job commitment and better staff performance and can raise prospects for sustainability.

Set Fewer Donor Restrictions on Hiring

The way things work now, technical cooperation projects are conceived, designed, and implemented mainly by external donors, with greater or lesser degrees of technical ministry collaboration and very little coordination or vetting by planning agencies or finance ministries. This is not everywhere true, and in almost all countries there are exceptions. Some countries (Guinea and Swaziland, for example) are exercising greater coordinating authority, as noted above. But these cases are only a small share of the total.

Donor dominance and the limits they put on recipient country hiring options contribute to many of the deficiencies noted earlier—the difficulty of establishing local control or coordination with the result that technical cooperation priority-setting is done mainly by donors; the tendency to overemploy long-term resident expatriates; the strong donor propensity to implement through donor country institutions such as universities, consulting firms, or nongovernmental organizations; and, especially, the inability to deal with that irksome paradox—that local competence cannot be hired because of budgetary austerity while foreign experts are financed by donors.

Almost all donors operate on the principle that aid resources should not be used for the financing of local costs. The principle is deep rooted. It derives from the basic rationale for foreign assistance—the removal of foreign exchange bottlenecks to acquisition of technology (embodied in capital goods) and skill. The pragmatic objection to local cost financing is also strong and is based on the belief that it is often not sustainable.

Despite their objections in principle, and their reluctance in practice, donors throughout Sub-Saharan Africa have financed local costs in the past, and have done so increasingly in the 1980s. But resistance to paying local salary costs remains strong and general. The issues in contention are discussed at length in the next chapter. We argue there that several good analytic and practical justifications exist for donor financing of local salaries in some circumstances: the minimization of project costs, relief of fiscal austerity that prevents use of locally available skills, and greater capacity-building impacts. In the present context, this means that formal and informal restrictions on hiring nationals for project design and implementation should be reduced or eliminated.

This is not an open-ended endorsement of local salary payments. As is also argued in Chapter Six, the practical disadvantages of donor payment of local salaries are enormous and they should be used only under circumscribed conditions. The acceptability of some local salary support in the present context is based on the inherently time-bound nature of true projects, making the sustainability issue less pressing.

The hiring of nationals for design and implementation on development projects should not founder on the recurrent cost issue—the continuation of the activity and the future absorption of those hired into the civil service. The underlying principle has to be that national technical assistance personnel should be financed only on true projects—that is, activities with well-defined beginnings and ends, and with objectives that are attainable in a reasonable time period.

The model should be a road project. When the road is designed, the consulting engineer leaves. When a maintenance system has been set up, the technical cooperation-provided expert who helped set it up goes. When the financial management improvement project in the power sector is over, local power company staff should be able to carry on by itself; the expert provided by technical cooperation (whether local or expatriate) should no longer be needed. This is the ideal: when the project is over, so is the employment contract of the technical cooperation involved.

The proposal, then, is to eliminate artificial segmentation of the market for technical assistance personnel by allowing and encouraging donors to use local consultants and consulting firms in project design work, and to hire more local experts in place of resident expatriates for project implementation when appropriate local skills exist. More generally, the proposal is to allow local actors a greater role in determining who to employ in designing and implementing projects that are externally financed.

African governments could do much to move donors in this direction by persuasion and by more focused and intensive dialogue on the issue. The effectiveness of this dialogue would be increased if more were known about existing donor policies and practices with regard to employment of nationals, and if the issues were clearly defined. We stress in the next chapter the need for studies of the salary supplements issue.

It would also help if technical cooperation programmes could be put in place everywhere. Dialogue with donors could then take place in the framework of the programme. This would be the forum for discussion of the input mix in all projects—capital or investment projects as well as technical cooperation. This is the advantage of a three-year programming cycle: governments could debate with donors, at an early stage in the project cycle, who should do the preparation or design, and who (nationals or expatriates) could and should be used for implementation.

In implementing these approaches, African governments can benefit from approaches that have emerged in Eastern Europe's experience with technical cooperation. In Poland and Hungary, the new governments and outside donors have had to devise vehicles for technical cooperation delivery that are at once simple, flexible, and efficient and that leave programming initiatives with the recipient governments. One such vehicle is the British Know How Fund and similar funds financed by other donors such as the French government. These funds are open lines of credit that ministries can draw upon for technical assistance of their choice. A local representative of the fund receives requests and helps find appropriate technical assistance.

The general idea is suggestive for Africa, although in Eastern Europe it is used only for expatriate technical cooperation. In African countries, a technical assistance fund could be created into which each participating donor would contribute. These contributions would be in the nature of lines of credit for technical cooperation projects. They would preferably be untied, but bilateral donors are likely to insist on using their own nationals if expatriates are used and this presents no major inconvenience.

Donor contributions to the fund could be based on estimated costs of the technical cooperation required for project implementation in the first year of the rolling technical cooperation programme—for projects, that is, that become operational that year. Each donor would transfer amounts necessary to cover the anticipated technical cooperation costs of the projects they will be financing in the first year of the technical cooperation programme and the public investment programme. The technical assistance fund would be the aid-financed equivalent of the technical assistance budget, which consists of the domestic contributions for payment of technical cooperation—the "fees" that are attached to each person-month of technical cooperation.

The central idea is that the resources allocated by donors for technical cooperation would be made entirely fungible insofar as who is employed on project design or implementation. The fund could finance local or foreign consultants and on whatever terms are appropriate, this decision being in the hands of the implementing agency and the national planning and financial authorities.

Consultation with donor agencies would have to be part of this approach, because donors who contribute to the technical assistance fund are in effect giving general budget and balance of payments support, and they need to be assured that these resources are being effectively used. Approval of the first-year technical cooperation programme by joint donor-government commissions should therefore be a requirement of release of technical assistance fund monies. That the appropriate fee has been put into the technical assistance budget would be another requirement.

The secretariat of the fund might include representatives of participating donor agencies, who would help user agencies find appropriate consultants and advisers by providing short lists, directories of consultants, and so forth. Donor representatives might also help in bid preparation and in suggesting donor country teaming arrangements for local consultant firms. Donor representatives could participate in contract awards and monitor implementation.

In addition to facilitating the reduction of technical cooperation market segmentation, the fund has a potential role in easing the allocation of aid funds to nongovernmental organizations. Nothing would prevent its taking on this role, should this be an objective of government and donor policy.

The proposal needs much fleshing out. Numerous accommodations would have to be made to country-specific conditions. But the basic notion is not complicated. The financing of technical cooperation by external donors should be separated from the design and implementation of these projects. The creation of a technical assistance fund will not result in complete nationalization of the technical cooperation process; donors will continue to have major roles through their continuing collaboration with user agencies. And the technical cooperation components of capital assistance or other types of development assistance will continue to be designed and financed as they are now. But the fund concept would, at the least, make it easier for recipient governments to have a voice in determining what kinds of technical assistance personnel are employed during the project cycle. And the fund concept will remove a significant distortion from the technical cooperation market.

PITFALLS TO THE MARKET APPROACH

Even these simple and logical steps toward rationalizing the use of technical cooperation would face numerous obstacles.

- Most donors would be wary, especially on the general issue of paying salaries to nationals employed on development projects.
  — Donors worry about the disruptive effects of dual salary structures—paying nationals on projects much more than civil servants.
  — Donors are especially concerned that empowerment of recipient governments, such as is implied in allowing them a greater role in the determination of project staff, might be premature in much of the continent. Its efficacy, for example, depends on the existence of a reasonably well-functioning technical cooperation programming system, which exists only in a few places.
  — Administrative and political imperatives that call for detailed financial accounting and controls might not be met without expatriate technical assistance personnel. Without strong systems of financial control and much transparency, devolution of financial management is regarded as susceptible to corruption.
  — Some important technical cooperation providers are not convinced that the deficiencies of present arrangements are sufficiently serious to require drastic changes. And most donors are reluctant to confront potential opposition from stakeholders in the existing system—universities, consulting firms, and some aid agency officials.
- The numerous technical cooperation projects would have to be carefully costed and programmed over the period of the rolling budget—usually three years. This is often not yet the case even for sizable capital projects.
- Because of staff shortages, it often makes sense for overworked planning agencies to concentrate their evaluation energies on the biggest, most important projects. If technical cooperation programming were to be included administratively in the public investment programme process, there is some risk that they would be neglected because technical cooperation projects are numerous and small.

- Technical cooperation projects would be exposed to the same risk now run by investment projects—that the domestic resources appropriated for local counterpart contributions to externally funded projects would not be allocated. These appropriations are often cut when budget shortfalls exist. If donors overlook these failures to deliver counterpart funding, as they often do in investment projects, the credibility of the exercise would quickly be compromised. But, at the same time, the technical cooperation environment would be much disturbed were donors actually to sanction nonpayment of the promised counterpart funds by delay or cancellation of scheduled technical cooperation projects.
- The monitoring of disbursements would be difficult because so much of technical assistance personnel expenditures takes place outside domestic financial circuits—payments of salaries directly into offshore accounts, for example, and offshore purchase of equipment and supplies. This would introduce especially large uncertainties into the budget process because the amount of last year's offshore spending on ongoing projects would not be known when this year's budget is drawn up.
- Adoption of this idea would create an anomaly. Many investment projects are financed 100 percent from foreign aid in all African countries. Under the proposed arrangements for technical cooperation, no technical cooperation projects would be so financed; all would have a domestic cost counterpart contribution.

These are only some of the potential pitfalls. More would appear as the idea was applied in different country circumstances. But what is most to the point is that the introduction of user fees and the setting of ceilings on technical cooperation expenditures would set in motion a process that will almost surely rationalize technical cooperation use. For example, user fees and ceilings would reinforce the effort to programme and budget technical cooperation. It would increase pressures on user ministries to cost and programme their planned technical cooperation properly. The shifting of payment burdens directly to user agencies would encourage greater transparency of costing and simplified, more efficient systems of payment. For example, there would be great advantage in shifting to cash allowance systems for items such as housing provided to personnel.

Because of the uncertainties and risks involved in such an innovation, the transition should be gradual and some changes tentative. For example, it might be desirable to start with a few key sectors or ministries, extending the process over time. Also, although it would be desirable in principle to include all technical cooperation personnel in the technical assistance budget, it might be prudent to start only with free-standing technical assistance personnel because of the difficulty of capturing all project-related technical assistance.

It would also be advisable to have substantial contingency lines in the first few years, as experience is gained with better costing and programming of technical cooperation projects. It would be unwise, also, to insist at the outset on rigid aggregate ceilings for technical cooperation because rates of disbursement will be highly uncertain; it makes little sense to struggle over the determination of such ceilings if experience reveals that year after year actuals are much lower than budgeted expenditures.[13]

This proposal, then, is to extend the concept of the recently introduced technical cooperation programme by creating for technical cooperation a mirror image of the currently popular rolling investment programming system, and to enhance it with an obligatory cost sharing out of domestic resources. To this is added the proposal for a technical assistance fund through which donor contributions to technical cooperation would be channeled. National authorities would have greater freedom to seek training assistance or help with the organization of supply and support systems, without necessary ties to technical assistance personnel.

We have focused on two points in this chapter. First, the market for technical assistance personnel is awry. It is a market with unseen prices in which supply and demand operate largely independently of cost considerations. Governments believe they incur few or no costs when they ask for or accept technical cooperation that is grant financed (as most is). And regardless of how technical cooperation is financed, the host agency (the user) has no incentive to economize on its use. To the contrary, the incentives run the other way because users pay nothing but enjoy many benefits from it. On the supply side, donors have many reasons to urge technical assistance personnel on host governments and few to abstain. So the incentive structure within which technical cooperation transactions take place is all wrong. This is a main reason why technical cooperation projects proliferate, local commitment is often minimal, and the use of technical assistance personnel is often out of line with national priorities. The proper policy response is to attach a price to the use of technical cooperation.

Second, the tied technical cooperation package of expatriate technical assistance personnel and equipment, supplies, and training also induces irrational use of these resources, and should be untied. Restrictions on hiring nationals for project design and implementation should also be reduced.

NOTES

1. OECD, DAC, "Principles for New Orientations in Technical Cooperation," Paris, 1991, p. 1.

2. European Center for Development Policy Management, "New Avenues for Technical Cooperation in Sub-Saharan Africa," October 1991.

3. World Bank, "Managing Technical Assistance in the 1990's: Report of the Technical Assistance Review Task Force," 1991, Annex: "The Technical Assistance Market and the Major Suppliers."

4. Several other kinds of costs are referred to in the text. "Transaction costs" are resources given up in negotiation and management of technical cooperation projects. They are real (opportunity) costs of a specific type. "Nuisance" costs in the text refers to costs absorbed in administering technical assistance personnel who prove unsuitable. They are one type of transaction costs.

5. Table 7.7 in the Statistical Annex shows that technical cooperation is between 25 and 50 percent of total official development assistance in two-thirds of the countries of the region. In only four countries is technical cooperation less than 20 percent.

6. In some cases, locally paid costs of technical cooperation are significant. In Sao Tomé and Principe, for example, many of the 122 technical assistance personnel were paid half their housing costs in addition to other local costs. And the Gabon government paid, at least until very recently, all the salaries of the 1,733 expatriate technical cooperation staff in its civil service. Sao Tomé and Principe, Round Table Conference document, vol. II, ch. IV, "Technical Assistance: Situation and Perspectives"; and UNDP, Mission Report of A. Barbesa, May 1992.

7. Several bilateral donor spokespersons say, when questioned on this point, that they operate with a global country and budget without predetermined proportions for the different types of aid. This is the case, for example, with the Netherlands aid agency and with A.I.D.

8. The nuisance costs are likely to be minimal because user agencies unhappy with an expert given under a technical cooperation grant can assign him or her to a routine task, or none at all. Benign neglect can thus sharply reduce the real costs represented by diverted management energies.

9. It should be noted, however, that local actors at all levels are not at all amenable to paying for technical cooperation, even a little. Governments everywhere in Africa share a common policy impulse: they try to substitute grant-financed for loan-financed technical cooperation whenever possible. This is a common theme in the national technical cooperation policy statements that have been prepared under NaTCAP auspices.

10. See Derek W. Brinkerhoff, "Institutional Analysis and Institutional Development: A Survey of World Bank Project Experience," CECPS, World Bank, October 1989.

11. As noted earlier, however, consulting firms—to win contracts—have some incentive to propose less costly personnel because cost becomes a major factor when aid agencies choose between proposals of equal quality.

12. World Bank, "Managing Technical Assistance in the 1990's . . . ," 1991.

13. This is what happens now in the public investment programme process in many countries: the size of investment programmes is hotly debated as a major macroeconomic issue, although every year the programme is substantially underspent.

6

GETTING THE WORK ENVIRONMENT RIGHT

Better planning and management of technical cooperation by African governments, more orderly and flexible technical cooperation provision by donors, and introduction of more market-like conditions into the technical cooperation process would all lead to improvements in the performance of this form of aid. But two critical obstacles would still have to be dealt with before technical cooperation could become a genuinely effective instrument for capacity building. The first is the severe organizational weakness that characterizes the public sectors of most African states. The second, closely related, is low public sector pay and the resulting sparsity of incentives for job commitment, skill acquisition, and hard work.

Some kinds of technical cooperation can have positive impacts despite the fact that they take place within weak and disorderly administrative settings in which local employees have minimal financial incentives. This is the case, for example, of gap-filling types of technical assistance, and of hard or engineering technical assistance related to capital projects. In these and related cases, the output is a specific, time-bounded product—for example, a study or report. Typical examples are the updating of existing civil aviation regulations and incorporating recent international conventions in them by an expert in civil aviation, and writing a report on local energy policy and defining possible solar energy strategies by a short-term specialist in solar energy.[1]

These kinds of technical assistance can augment the stock of local knowledge, lead to development programmes that are better designed and executed, and improve public policy making and management, whatever the nature of the administrative environment.[2] They do so by operating more or less autonomously, making few demands on local institutions. They can, within limits, get a job done. But because they typically transfer few skills to local people and have little gearing or engagement with local institutions, they do extremely little to build local capacity.

To be an effective vehicle for capacity building, technical cooperation requires a minimally congenial administrative environment and motivated, stable public employees. The absence of these preconditions dulls the impact of even well-conceived and well-executed technical cooperation projects. Until recently, this uncomfortable reality was not much taken into account in technical cooperation policies and programmes. Yet it has long been evident that without an appropriate work environment, and especially without local staff commitment and stability, technical cooperation fails to "take" or "stick"; it operates in a Teflon-like world in which the beneficial impacts of institution-building efforts slide away over time.

Thus, right from the beginning after independence, several typical features of African work environments diminished the impact of the capacity-building technical cooperation that was tried. Stocks of skilled and educated African manpower were small, demand for this labor intense, and increases in supply relatively slow. Until about 1980, in most countries labor market conditions remained favorable for skilled local people because of expansion of public services and the push for replacement of expatriates.

In this context, labor market conditions in the internal public sector encouraged high job turnover. It did not pay competent and ambitious employees to stay in any one job. If they did, they would probably receive only gradual salary increases via movement up

Incentives for high staff turnover

incremental scales and slow promotion. The best strategy for success was to change posts often and to move from technical to administrative assignments, in which prestige and money rewards are higher.

This kind of setting created the wrong incentives for skill acquisition and institutional strengthening via technical cooperation. Effective skill and technology transfers require job-specific training and job stability. However, the career interests of local staff counterparts to technical assistance personnel were served less by deepening of technical skills, which are often job specific, than by acquisition of more general abilities. (It is noteworthy that the kinds of skills best transmitted via the counterpart relationship are technical and job specific.) Nor did local staff have as much interest in in-service and on-the-job training as in broader training and in diploma-yielding study, especially at universities in developed countries. The latter provide better training and surer credentials for mobility.

An incentive structure that favors acquisition of general skills and much interagency job mobility provides infertile ground for technical cooperation and is not conducive to institution building in general. Combined with other factors favorable to short-time horizons (frequent ministerial changes, nepotism in promotion policy), this structure weakens internal impulses to build public sector institutions from the ground up.

A second factor related to the administrative environment also diluted the impact of technical cooperation on capacity building. The capacity of public sector organizations depends on more than the level of skills and technology at their command. Capacity is determined by the efficacy of administrative operations—personnel, financial, training, and supply management systems. It is also conditioned by external factors such as growth in gross domestic product and availability of budget resources. Although technical cooperation was working to introduce new technology, improve skills, and strengthen institutional procedures, most components of the administrative system, and of the general economic and financial environment within which the administrative system operated, were deteriorating throughout the continent.

For many years, then, unfavorable elements in the work environment have reduced the capacity-building impact of technical cooperation. And in the past decade these elements have become more pervasive, as the rate of decline of real salaries accelerated in many countries and as budget austerity took a toll on the effectiveness of government operations.

These considerations explain the urgency of the central question addressed in this chapter: how to create work environments that are favorable to capacity building and thus will enable capacity-building technical cooperation to be more effective. The chapter has three parts.

- The first part describes the deficiencies of administrative organization and the erosion of wage incentives that have become endemic in Sub-Saharan Africa.
- The second part summarizes donor and African government responses:
  - The short-term response has been to expand greatly the practice of paying local salaries and salary supplements. Almost all donors take the position that salary supplements, like all external financing of local costs, are bad in principle and have negative side effects. But salary supplements are often necessary in practice. Their effects, however, are to disrupt the work environment because of induced job turnover and the resentment of civil servants who do not receive supplements;
  - Major effort has gone into civil service reform and, in particular, reform of government pay and employment policies. A review of evaluations that have assessed the results indicates that impacts have been few; and
  - Some attempts have been made to undertake more general administrative reform, but these do not seem to have been numerous or highly intensive. Little is known about their results. The lessons of administrative reform experiments in general are not encouraging.
- The final section addresses the question: what is to be done to improve work environments so that they are more congenial to capacity building, and thereby more responsive to technical cooperation capacity-building efforts. With respect to salary supplements and external financing of national employees, we

propose that nationals be hired only for project-related jobs and for projects in their true sense of activities with a beginning and an end. We propose intensified study of the supplements problem and much more open, informal coordination among donors in-country.

Expectations have to be modest about the results from administrative reform efforts in general and salary reform in particular. These are notoriously difficult, slow, and uncertain. Although reform efforts should be intensified, it would be imprudent to ignore the high risks of failure and the long time that will be necessary to transform administrative environments. A parallel strategy of building capacity outside the public sector is therefore recommended on the grounds that it may not be possible to make significant improvements in public sector work environments in the medium term—the next 10 to 20 years. This means seeking alternative ways to provide vital public services, notably by accelerating the processes of deregulation and privatization already under way in most of the continent.

PUBLIC SECTOR ORGANIZATIONAL DISARRAY AND LOW SALARIES

Public sectors in most of Africa have experienced a downward spiral in operating efficiency in the past 20 years. This decline is caused partly by deterioration of basic management systems and partly by inadequacies in salary and structure.

Public Management Deficiencies

The major operating shortcomings of civil services in Sub-Saharan Africa are distressingly familiar. For reasons outlined earlier, administrative efficiency has deteriorated on a wide front. The nature of this deterioration is summarized graphically in a recent report of the U.K. Overseas Development Administration, on which we draw in the following discussion.[3] It should be noted that the catalogue of problems that follows, although severe and widespread, is hardly unique to Sub-Saharan Africa and does not afflict all Sub-Saharan African countries equally.

The major problems in civil services are now well documented.

- There are too many civil servants—more than are needed to do their assigned work. This assertion (which is difficult to prove in a rigorous way) is based on the fact that employment from the mid-1970s tended to grow faster than population and nonsalary budgets, on casual observation of the operations of government departments and services, and on the results of numerous management audits of individual agencies.
- Civil servants are poorly organized, graded, and paid. Real wages for skilled technical people and managers have fallen so low that they no longer provide "traditional" middle class living standards. Wages of higher-level staff are also low relative to parastatal and private employees, and compared with wages of lesser-skilled civil servants. Civil servants are therefore poorly motivated, and most work ineffectively.
- Despite individual pay levels and allowances that provide tenuous support for employees and their families, civil services cost governments far too much for national revenue-raising capacities. Salary costs eat up the essential nonsalary expenditures required to make civil servants effective.
- Revenue generation systems, notably in income tax and in customs, excise, and sales taxes, have declined and in some cases collapsed. Decline is not just due to circumvention and corruption but to the inability of the government departments responsible to operate their systems. Revenue shortfalls are often unexpected and unexplained, and play havoc with government budgeting and cash management.
- Difficulties in budgeting, financial planning, and monitoring have led to symbolic budgets, unconnected with actual funds available or with stated priorities; to recurrent liquidity crises; and to major problems in allocating aid and foreign investment and in managing external debt. Planning agencies no longer play allocative or forward planning roles (if they ever did), and

become mainly compilers of lists of sectoral development projects and channels for the transmission of aid funds. Expenditure controls have weakened, budget estimates have become increasingly unrealistic, and late production of government accounts has become more pervasive.
- Budget systems pose additional obstacles to good organizational performance. They are frequently rigid: requests for new money for completed projects are rarely granted, and requests for transfers and other changes yield slow and uncertain responses. Programmes are poorly costed and often inadequately budgeted. Line item budget systems prevail and provide little transparency. Budgets are often not comprehensive; much spending goes on outside the budget process. It is often unclear how uncommitted revenues are allocated between investment and operating budgets and among technical ministries.
- Personnel management is inefficient and often inequitable. Selection and appointment procedures are cumbersome. Employment record keeping is often inadequate. "Phantoms" often populate the payrolls—people long dead or never alive whose paychecks nonetheless are regularly paid and cashed. Hiring and promotion are not always based on merit. Disciplinary systems are soft; nonperformance rarely leads to dismissal, absenteeism is overlooked, and incompetent staff are protected. Grading systems and promotional criteria create no incentives to improve performance—positions and salary grades are too close together, and allowances are sometimes so large relative to salaries that they introduce distortions in work patterns (for example, traveling expenses). Grievances go unheard and fester because trade unions or other consultative arrangements have atrophied and in-house grievance handling systems are nonexistent or rudimentary.
- Civil services have been assigned so many developmental and welfare responsibilities that they have far outrun the resources at government disposal. The growing reality is that civil services offer declining real benefits to the average citizen and operate to some extent symbolically. They also function partly

as social welfare and employment creation agencies for the educated middle class. Even this role is now weakening as school and university output grows and public money for new jobs dries up.

- There are fundamental problems of internal management and efficiency. Weak controls on finance are matched by similar weakness in controls over supplies and equipment; waste and corruption have become more commonplace. Maintenance is rarely done. Secretarial and registry services have broken down—letters can neither be sent nor stored in usable information systems. Basic bureaucratic requirements such as organization charts, job descriptions, operating manuals, and standing orders either are out of date or have disappeared.

- Vital training systems have broken down so that, for example, essential support staff in bookkeeping, storekeeping, and registry work are not trained. In-service training programmes are few in number and often of limited quality. Specialist groups such as customs and excise personnel and government accountants learn inadequately on the job because experienced trained supervisors are few and far between. In part as a result, career development concerns usually receive insufficient attention. Responsibility is begrudgingly delegated. Scholarships for advanced study are not abundant and are found as part of foreign aid projects.

- In some countries, high ministerial turnover is a serious problem. Ministers change frequently and with them key management staff, often at the director level. Plans carefully laid have to be reworked, often abandoned; approaches once in favor fall by the side; relationships once cemented become irrelevant.

These dreary symptoms of organizational disarray are more prevalent in some countries (Ghana, Sierra Leone, Chad, Sudan, Uganda, Tanzania, Mozambique, and Zaire, for example) than in others (Zimbabwe, Kenya, Senegal, and Nigeria, among others). Overall, however, during the past decade only a minority of Sub-Saharan African countries—probably a small minority—have been able to avoid significant deterioration of already weak administrative capacities.

All of this points to the existence of a basic paradox. Technical cooperation for capacity building aims at improving organizational competence. But its effectiveness depends to an important extent on the existence on the ground of some degree of organizational cohesion and stability. When organizational disarray is profound, sustainable impacts are likely to be few, even when technical cooperation programmes are well planned and carefully designed, superbly staffed, and diligently supervised.

This suggests another important point. In this chapter and throughout the book, the focus is on salary policy and incentive structures. The other critical elements in administrative performance (such as capacity in revenue raising, budgeting financial management, training, nonsalary aspects of personnel management, and support systems) receive little or no comment. This is a clear imbalance. Inadequate wage incentives are a major source of administrative deterioration, but they are not the whole problem. Much institutional weakness would remain even if salary issues were less pressing than they have become.

The imbalance in treatment reflects the reality that donors and governments in Sub-Saharan Africa have until recently given top priority to personnel management (along with public investment programming) in their institution-building efforts. It is not intended to minimize the significance of other facets of the administrative reform problem. One point is clear: administrative systems struggling under the burdens of a deteriorating structure and absorbed in short-term crisis management are fertile ground for institution-building effort. Implications for the role of the state and for technical cooperation strategies are considered later.

Inadequate Salary Incentives

In the past two decades, most African governments pursued wage (and related employment) policies that had three common features. First, they favored employment growth over income growth of public employees, so public payrolls grew, whatever the rate of growth of the economy or public revenues. Second, they let real wages fall; in all but a few cases, public sector salaries failed to keep pace with inflation. Finally, in allocating pay increases, African governments invariably favored the lower-paid ranks. Real wages of skilled and educated staff fell more than those of unskilled workers.

Table 6.1 summarizes evidence illustrating these main policy themes. The last column shows that governments tended to follow aggressive public employment policies during this period, which was, in most cases, one of slow economic growth or stagnation. In making choices between higher salaries for existing staff and greater employment, governments opted for greater employment in almost all cases.

Although declines in real salary levels were quasi-general, cuts were much sharper for skilled and educated employees than for unskilled. In adjusting money wages to take account of inflation, governments systematically favored low wage jobs. As Table 6.1 shows, in 1983 real wage rates for those at the top of the civil service were 11 percent of what they were in the mid-1970s in Ghana, 5 percent in Uganda, 30 percent in Nigeria, and 45 percent in Zambia. Unskilled rates deteriorated much less.

Table 6.2 shows the same phenomenon for a largely different set of countries, and with data from a different study. The base and terminal years are also different. The story told by this table varies in some respects from that in Table 6.1; the relative decline of skilled worker wages seems less marked. This difference suggests that the extent of the decline in real skilled wages is not so firmly known. But that it occurred, was often sharp, and affected higher-paid people most is confirmed in the data in Table 6.2.

These data are for rates. Earnings data are more difficult to find. Generally, average earnings series (government wage bills divided by number of employees) tend to show sharp reductions during the 1980s—40 percent is not an uncommon rate of decline over the decade. Factors other than reductions in individual real salaries can explain these declines—a shift in composition of the government work force toward less skilled employees, for example.

TABLE 6.1

WAGE AND EMPLOYMENT CHANGES, 1975-1983,
SELECTED COUNTRIES

Country	Index of Real Wage Rates 1983 (1975-1977 = 100)		Public Sector Employment % Increase per Year
	Unskilled[a]	Highly Skilled[b]	
Ghana	40	11	+15
Malawi	85	65	+ 8
Nigeria	64	30	+15
Senegal	112	74	+ 5
Sierra Leone	103[c]	62	+ 6
Sudan	35	29	+ 3
Uganda	15	5	-
Zambia	88	45	+ 2

Source: Barbara Nunberg and John Nellis, with Louis de Merode, "Civil Service Reform and the World Bank," World Bank, December 1989.
[a] Messenger or lowest beginning rate, bottom scale.
[b] In most cases university graduate starting rate; otherwise, principal secretary.
[c] 1980.

TABLE 6.2

1985 BASE CIVIL SERVICE SALARY RATE AS PERCENTAGE OF 1975 RATE

Country	Lowest Grade	Highest Grade
Benin	43.2	43.2
CAR	48.9	35.4
Ethiopia	62.9	31.3
Gambia	39.5	36.2
Kenya	58.2	42.0
Mauritania	60.6	43.8
Morocco	66.4	52.4
Niger	46.2	46.2
Nigeria	42.2	21.9
Sierra Leone	23.4	15.7
Somalia	5.2	4.0
Sudan	30.3	24.6
Tanzania	22.9	18.8
Togo	58.1	58.1
Zimbabwe	150.7	58.1

Source: D. Robinson, *Civil Service Pay in Africa*, ILO, 1990

Individual earnings have almost certainly not fallen by as much as indicated by the change in salary scales or rates.[4] Civil servants receive incremental increases on a regular basis, and many are promoted. Also, nonsalary compensation elements tend to increase when salary scales are raised less than the rate of inflation. So wage deterioration in many public sectors has been less than is indicated by the salary scale data used in most comparisons. But there is no doubt that, except in a few countries, real salaries have fallen significantly both in entry rates for different jobs and in average earnings of established civil servants. Government salary levels at the beginning of the 1990s were so low that many civil servants viewed them as inadequate to meet customary needs.[5]

Another wage-related factor contributing to a weak incentive system within many public administrations is the growing divorce between effort or responsibility and reward: the amount of salary received by employees is less and less determined by the importance of their contribution to the functioning of the public sector. This has two causes.

- First, the egalitarian (or politically driven) salary policies of the past two decades have resulted in much-narrowed skill or responsibility differentials. The data are weak and allow conflicting interpretations, although this difference appears to be due mainly to differences in definitions of who and what is being compared. But the evidence (summarized in Table 6.3) leaves little doubt that rewards for responsibility and skill, traditionally very large in African countries, have generally diminished, in many cases radically. Some case study evidence, using closer definitions, finds greater salary compression. In Ghana, for example, average pay for highly responsible jobs at the upper levels of the civil service was only two to three times as high as average pay in the lower skill categories in the late 1980s.[6]
- Second, nonsalary benefits seem to have grown as a share of total compensation. Evidence is sparse, but some case studies do exist. Salaries as a percentage of total compensation in the Gambian civil service fell from 88 percent in 1982 to 78 percent

TABLE 6.3

SKILL DIFFERENTIAL RATIOS FOR SELECTED COUNTRIES
(ratio of highest to lowest wage rate in civil service)

Country	Year	Ratio of Highest to Lowest Rate
Burundi	1984	17.2 : 1
Cameroon	1989	22.0 : 1
CAR	1985	9.4 : 1
	1988	9.0 : 1
Gambia	1985	8.5 : 1
	1988	5.7 : 1
Ghana	1984	5.7 : 1
	1989	7.8 : 1
Guinea	1985	8.7 : 1
	1987	3.8 : 1
	1988	4.5 : 1
	1989	4.6 : 1
Guinea-Bissau	1988	5.2 : 1
	1989	4.0 : 1
Mali	1985	16.5 : 1
Mauritania	1975	6.9 : 1
	1985	3.0 : 1
Niger	1975	18.2 : 1
	1985	14.8 : 1
Senegal	1980	7.8 : 1
	1982	7.3 : 1
	1983	6.8 : 1
	1985	6.5 : 1
Togo	1985	12.0 : 1
Uganda	1983-1984	5.9 : 1
Zaire	1985	47.2 : 1

Source: Barbara Nunberg et al., 1989.

in 1988; in Senegal, from 75 percent in the early 1980s to 57 percent in 1989; and, in the Central African Republic, from 62 to 59 percent between 1984 and 1987.[7]

These salary trends have profound implications for capacity-building technical cooperation. The shrinking of real wages, and especially the wages of the skilled and educated employees, has led to widespread discontent on the job, low commitment to public sector work, much moonlighting, some corruption, and high staff turnover—in sum, a withdrawal of effort and diminished involvement at the work place.

Technical cooperation projects operating in these environments suffer the same fate as other activities. Local staff have little financial incentive to work as counterparts, to work hard in counterpart relationships, or to invest time and energy in job-specific training of the kind technical cooperation provides. As noted earlier, the incentives for local staff are mainly to acquire general, transferable education and training to increase their potential mobility. On the job, they are distracted by the need to earn supplementary income. Often, they succeed in finding other posts inside or outside government that offer better opportunities.

Commitment to the project, and especially to its institution-building dimensions, is thus usually slight. It is no wonder that a merry-go-round of counterpart search and replacement bedevils so much technical cooperation effort, or that the Teflon analogy comes so easily to mind in assessing its capacity-building effects.

DONOR AND GOVERNMENT RESPONSES

Three lines of attack are discernable.
- A short-term, crisis-management response has become generalized in recent years: donor payment of direct and indirect salary supplements to local employees. African governments do not discourage these practices and in most cases openly encourage them. The main purpose is to recruit or retain local staff for donor-financed projects. In some cases, salary costs for whole groups of local employees are paid by donors.

- With strong encouragement from donors, many governments have introduced salary and related employment policy reforms. These aim at slimmed-down, better-paid civil services, with much better wages for the highly skilled. The cost containment aspects of these reforms are often linked to the expenditure reduction aims of IMF and World Bank adjustment operations.
- Some donors and governments have given priority to broader reforms of the civil service or, still more broadly, have addressed the issue of general administrative reform.

The nature and outcome of each of these approaches is considered in turn. Implications for technical cooperation are analyzed in a later section.

Donor Payment of Local Salaries

Deteriorating wage incentives in the public service, combined with new conditions in African labor markets (unemployed university graduates), have created strong incentives for donors to supplement local salaries and to hire local staff.[8] The official position of almost all donors has always been to oppose payment of local costs and of local salaries (including supplements) in particular. Yet almost all pay them. Like partner African governments, which introduce various cash and noncash benefits and hire contractuals on vastly better terms than civil servants, donors seek to escape from the constraints imposed by civil service salary policies.

The issues are complex; it is not easy to know even what kinds of supplements are paid and casual discussions are often unclear about the kinds of employees in question. Few studies of the problem have been done.

It helps first to sort out who are the employees in question. They are mostly skilled and educated staff—the higher grades of the civil service. They are hired mostly for employment on donor-financed projects. Five specific types can be distinguished.
- The coordinators, administrators, or national directors of development projects. These are often detached civil servants. Their total salaries and benefits are paid by project budgets.

- Working staff of development projects. Often these employees are paid entirely from the project budget. The standard model would be the hiring of an expert—a national specialist in seed multiplication or plant genetics, for example—by a seed multiplication project. There are many variants of this model:
  — Supporting staff working in a bureau that houses a technical cooperation project may be put wholly on project payroll. In Rwanda, for example, an agricultural statistics project pays full salary costs of several data entry staff, secretaries, and research assistants;
  — In some institution-building projects, professional staff are hired on contract and paid by project funds. In Senegal, a project to strengthen development management pays the salaries of most of the higher-level professionals in the planning agency, who are hired on contract; and
  — In Uganda's Ministry of Planning, almost all staff gets some monthly topping-up payment from a planning assistance project, and even staff in departments unrelated to the project are given supplements.
- Then there are formally designated counterparts to resident expatriate technical assistance personnel. These key personnel are often given supplements to encourage their involvement in the project. In some cases, national experts are financed by donors to act as counterparts.
- Blanket topping-up is not uncommon: supplements are given to all or most employees in an administrative division or department that has a technical cooperation project. Sometimes these spill over into related departments.
- Local consultants are increasingly hired with donor resources to do studies and research that were formerly the monopoly of donor country consulting firms and universities.

Numerous vehicles have been devised by which supplementary payments can be delivered. Payment of full salaries for project personnel was mentioned earlier. Topping-up of counterpart salaries is also done, although indirect rewards—bonuses for overtime or special work assignments, car allowances, and per diem payments are probably more common. Some donors use the approach of hiring civil servants as consultants for a few weeks (during vacation, for example) or paying them to produce a specified paper.

It is commonplace for project-related staff to receive special allowances for cars and gasoline and per diem for local travel. These can be highly remunerative; it is not unusual for four or five days of local per diem to exceed monthly salaries. Per diem with related travel to foreign conferences or seminars is another favorite. Non-personal-income benefits are also attractions: a decent office, access to secretarial help and office equipment (copiers, computers), access to project cars, and promise of fellowships for foreign travel, study, and conferences.

When qualified local people are available, recruitment of nationals to project posts is cheaper and usually better than recruitment of expatriate personnel. Local recruitment permits implementation of projects at lower cost and is more effective in developing national capacity because it facilitates internalization of the technologies, procedures, and skills being transferred by the project. But recruitment of nationals also carries severe disadvantages: a disorderly salary structure, greater job turnover, lack of transparency in remuneration, discontent among those left out, increased dependence. Payments of salary supplements have the same effects. We consider the advantages and disadvantages more systematically in a later section.

No donor openly endorses salary supplements, and almost all are in principle against the financing of local staff salaries with aid money. Most try to minimize the potential damage of these measures and put restrictions on them.[9] But salaries of civil servants are low, moonlighting to make ends meet is frequent, and grievances with general conditions of work are widespread. So donor staff responsible for project implementation argue that the elimination of wage supplements would cripple all development activity. Without them, they say, the development projects they finance could not find local staff and counterparts. The cost of their projects would be higher because more expatriates would have to be brought in. As long as some donors continue to pay supplements, all must do so to compete.

This explains the schizophrenia that surrounds this issue. One major donor responded as follows to a 1990 DAC questionnaire on salary supplements.

> The aim of every project is for [our] contribution to become superfluous as quickly as possible . . . so that the partners can assure their respective tasks without external assistance. Salary supplements can lead to dependence. . . . For these reasons [we] are careful to ensure that in principle the agreement is that the partner is responsible for all local staff costs.

The donor then notes, however, that supplements can be paid "if it is necessary in the interest of the project." The decision is taken at the project level.

Another donor agency asserts, in a paper written in response to the same questionnaire, that its policy is not to pay supplements.[10] When these payments are made, it is done without official sanction by project executing agencies.

Most donors find complex and imaginative formulas to finance incentives or salaries for national personnel who perform functions that would normally be done by civil servants or counterparts. Ad hoc benefits are often preferred because they are less visible and can be accommodated more easily to both recipient government and donor regulations. This explains the spread of practices such as holding seminars outside the capital city to justify payment of per diems, or awarding short-term consultancy contracts to civil servants for special assignments. These are less disturbing to existing salary structures than more formal fringe benefits, and can be less disturbing than hiring nationals for project work at many times the civil service rate.

It is important to point out that in these kinds of situations, when open and hidden wage supplements are provided to national employees and when African governments may ask for expatriate technical assistance personnel because they cannot afford to hire equally qualified local people, technical cooperation takes on a role different than its traditional one: it substitutes for and subsidizes government operating budgets. It does this directly by payments to government staff on projects, and indirectly by financing experts to do operational work normally done by government employees. This is disadvantageous in two ways. It misuses the technical assistance personnel resource, reducing its effectiveness for institution building. And it is extremely costly; high-cost expatriates are hired in posts that nationals could fill more cheaply.

The introduction of salary supplements by donors has created as many problems as it has resolved. It has probably led on balance to a worsened work environment and made the task of technical cooperation for capacity building more difficult. We consider later what might be done to reduce its negative effects. But first the results of efforts to address longer-term administrative obstacles are reviewed.

Pay and Employment Reform

The main effort for administrative reform in the 1980s focused on civil service employment and pay. The two were linked based on two assumptions: that overstaffing is endemic in African civil services and that it is fiscally possible to finance a salary structure with appropriate incentives only with a leaner civil service.

The biggest external actor in civil service reform and the chief architect of reform strategies is the World Bank. Between 1981 and 1989, the Bank financed 76 projects with significant civil service reform components, 47 of them in 20 Sub-Saharan Africa countries.[11]

The principal short-term objective of the World Bank approach is to cut back the level of employment, especially of lesser-skilled civil servants, who are generally plethoric, and to increase real wages of all civil service employees, but particularly those at the higher ends of the wage ladder—the managerial and technical groups most disaffected under present arrangements. What is sought is not only a substantial increase in real salaries at the top but also a substantial widening of pay differentials. This is needed to discourage moonlighting and encourage acquisition of skills and acceptance of responsibility. It is also needed to avoid fiscal disaster.

The main lines of attack on the excessive employment problem are well known by now: the elimination of phantoms from payrolls, posts that are vacant, discharge of temporary [staff], [enforcement] of legal retirement age regulations, hiring [freezes], [no] guaranteed entry to high school and university [graduates], [voluntary] departure (incentive) schemes, and outright [dismissals].

[...] in objectives of raising average real salaries [...] are to be achieved by changes in government [...] exercises, which redefine job content and [...] commonly used to re-establish order in civil [service which had be]en regraded for many years. They may have [the effect of restorin]g skill and responsibility differentials that [had been eroded over ti]me. In some cases also (Ghana, for example), [these element]s of the reform programmes are part of the [...]; the main instrument is government's periodic [budget].

[...] make possible some assessment of these reform [efforts. The employme]nt-reduction push has had some positive [results, as summ]arized for seven African countries in Ta[ble ...]. [Ghana and Ugand]a have made the most substantial reductions. [Ghana has cu]t its 1985 civil service employment by 16 [percent; Uganda has] cut 23 percent of its mid-1980s complement. These reductions are significant, but some care is needed in interpreting them. In all cases, control over new recruitment is not firm, so new hires have taken place. In Ghana, some observers estimate that new recruitment may have amounted to as much as 25 percent of the number of retrenchments, and that a significant number of these were in the same occupational categories as those retrenched; a few are people who were retrenched earlier.[13]

In the Guinea case, the payroll numbers are so poor that the reliability of the estimated shrinkage is low. Some of those discharged may still be receiving pay from a special fund. And nothing is known about uncontrolled hires.

The major purpose of retrenchment is not to slim down the civil service for its own sake, but to generate savings in the wage bill that will permit payment of higher wages to those who remain, and to allow more spending on nonsalary inputs. Here the results are discour-

TABLE 6.4

EMPLOYMENT REDUCTION MECHANISMS AND RESULTS FOR SELECTED AFRICAN COUNTRIES, 1981-1990
(number of employees affected)

Country	Ghost Removal	Enforced/Early Retirement	Voluntary Departure	Retrenchment (regular staff)	Retrenchment (temporary staff)	Other Mechanisms	Total[a]
Cameroon	5,830[b]	5,000	-	-	-	-	10,830
CAR	2,950[b]	-	1,200	350-400	-	-	4,500-4,550
Congo	-	-	-	-	-	2,848	2,848
Gambia	-	-	-	919	2,871	-	3,790
Ghana	11,000[d]	4,235[e]	-	44,375[e]	-	-	48,610
Guinea	1,091[f]	10,236	1,744	-	-	25,793[g]	38,864
Guineas Bissau	800[d]	945	1,960	921	-	-	3,826
Mali	-	-	-	-	-	-	600
Papua New Guinea	-	-	-	2,300	-	-	2,300
Sao Tome & Principe	-	-	4	-	294[h]	-	298
Senegal	497	747[i]	1,283	-	-	-	2,527
Uganda	20,000[j]	-	-	-	-	-	20,000

Source: World Bank, "The Reform of Public Sector Management, Lessons from Experience," Policy and Research Series no. 18, 1991.

[a] Gross figures not adjusted for new recruitment and attrition; [b] Elimination of ghosts and double payments; [c] Attrition through hiring freeze; [d] Refers only to ghosts identified. Their removal has not been verified in technical analysis; [e] Includes staff in district assemblies and in the education services; [f] Ghosts in Conakry, Guinea. A second census in 1990 identified a large number of additional ghosts; [g] Of this figure, 10,810 officials were assigned to a personnel bank and placed on administrative leave, and 14,983 were removed from civil service rolls. It is unclear whether all those placed on leave have left the service; [h] An undetermined but small portion of these may be regular staff; [i] An additional 2,123 officials have applied for voluntary departure or early retirement; [j] Estimate based on savings from ghost removal exercise divided by average civil service wages.

aging. In 15 countries (most of them African) for which wage bill data are available, and which had World Bank or IMF reform programmes, 10 had wage bill increases during the second half of the 1980s.[14] This group included countries that had made much progress in reform: Gambia, Ghana, Guinea, and Jamaica. Wage bills declined in only four countries. Moreover, wages as a share of total expenditures rose in nine countries, and the ratio of salaries to supplies worsened in seven countries; it improved in only four.

Data on real salary increases, and on salary decompression, are equally discouraging, as are those on changes in the remuneration mix that would link reward more closely with effort.

- Average real civil service pay has risen since the mid-1980s in a few countries. Guinea, where real wages rose by 120 percent between 1986 and 1989, seems to be the most striking example. But in Ghana the increase was negligible, and in Gambia real wages declined further. In most of the countries with civil service reform programmes, real wages seem to have changed little.

- The sparse data show no clear tendency toward widening of skill/responsibility differentials. In Ghana some decompression did occur; according to one calculation, the compression ratio there rose from 4:1 to 5:1 between 1986 and 1990.[15] But it was unchanged in the Central African Republic, Gambia, and Guinea, among other countries.

- From very sketchy evidence, it does not appear that the policy urged under the reform programmes to increase direct or base salaries and reduce allowances has had much impact. If anything, allowances have tended to become a bigger part of the salary package.

The spotty data that exist, then, suggest strongly that the impacts of the considerable reform efforts made to date in several African countries have been very slight. Employment has been cut, deeply in a few cases, but wage bills have not shrunk at all or not enough to finance salary rationalization. Real wages have risen on average since the mid-1980s in a few countries, but the key salary policy reform objectives—and in particular the goal of increasing salaries at the top more than at the bottom—have not been reached. Even in the countries in which progress in raising real wage levels has been significant—Guinea and Ghana, for example—the increases have not been big enough to cut moonlighting and other manifestations of loose commitment to the job.

The prospects that this will change do not seem bright. The assumption on which efforts to reform salary structures are based is that African governments will cut employment of low-skilled, low-paid people and hold down real wage increases of those who remain, while raising more rapidly the salaries of civil servants in the top grades. But few politicians, trade unionists, or intellectuals are likely to be receptive to a call for greater relative sacrifice for the more numerous and poorer employees at the lower ends of the civil service.

This call for a widening of skill differentials flies in the face of the egalitarian wage policies that have prevailed for 30 years in most of Africa. When making general salary adjustments to compensate for inflation, for example, African governments have almost always favored lower-paid employees. This policy thus has deep historical, ideological, and political roots.

Two recent examples illustrate the problem at hand. In Gambia, average real wages of civil servants declined by some 50 percent between 1980 and 1986, and surveys of comparative wages found private sector salaries on average almost 90 percent higher than those in the civil service. The differentials were larger at the higher skill levels. Following the appointment of a salaries commission in 1988, a pay increase averaging 67 percent was approved in January 1989. But the allocation of the increase was such that the ratio between high and low salaries was not widened. And this occurred in the face of an ongoing reform programme and much dialogue with the World Bank and IMF about the need for greater pay differentials for skill and responsibility.

The other story comes from outside of Africa. The World Bank's third SAL in Jamaica contained a condition that salaries of higher-level management posts in the civil service should be raised to 95 percent of the pay of equivalent grades in the parastatal sector. The objective was to stem a drain of managers to the parastatals. Civil service managers did receive two raises. But the third and crucial raise was blocked by the opposition of employees at the lower skill levels and their trade unions. Civil service salaries remain uncompetitive, and the attachment of civil servants to the job and their intensity of effort are said to be unchanged.

General Administrative Reform

The problem of civil service cost containment and rationalization of pay structures absorbed the bulk of reformer energies in the 1980s. Programmes with broader reform objectives were also undertaken, however. The U.N. agencies, especially UNDP and UN-DTCD (Department of Technology, Cooperation and Development) have financed numerous training and civil service improvement programmes. Bilateral donors have done the same. The French government, for example, has provided extensive assistance in financial management to many Francophone countries.

The U.K. Overseas Development Administration has been among the most active of the bilateral donors in this area. Its Public Administration Training and Institutional Development Programme in East Africa has projects in Kenya, Uganda, Tanzania, and Sudan. Some of these projects focus on training, others on institutional assessments as a first step in broader reform efforts. The Overseas Development Administration has also assisted in financing Civil Service Reform Commissions in Uganda and Zambia, and is active elsewhere in Southern Africa.[16]

The World Bank's activities in civil service reform (other than in pay and employment policy reform) have concentrated heavily on personnel management. They aim at stronger personnel information and management systems, more tightly linked to payrolls; staff audits; improved training; clearer legal frameworks for the civil service; and better general working conditions and recruitment policies.[17] The Bank has also financed (and executed for UNDP) many projects in public investment planning and programming.

We reviewed in Chapter One a large number of evaluations of these kinds of reform programmes, which all fall in the category of institutional development or capacity building. The evaluations reviewed invariably concluded that the record of institutional and administrative reform programmes is poor. One recent internal review by the World Bank's evaluation unit found that only 5 of 19 institutional development or public sector management projects could be judged even moderately successful.[18] A larger 1991 review of technical assistance for institutional development found that the bulk of this assistance failed to make significant impact and is of doubtful sustainability.[19]

The academic literature does not paint a brighter picture. Writers vary in the degree of diplomacy or politeness with which they describe results of past efforts in administrative reform in Africa—from "below expectations" to "dismal."[20] In this they follow in the tradition of general writing on the subject. Students of development management and administrative reform differ about why progress is so slow everywhere: some blame inappropriate reforms and others faulty implementation strategies. But everybody agrees that it is difficult, that every step forward is subject to high risks of slippage, that true success stories are few, and that in all cases it is a slow and long process.[21]

WHAT SHOULD BE DONE

Organizational disarray and lack of economic incentives in public employment thus continue to undermine institutional development in general and especially technical cooperation efforts for capacity building. The reform programmes of the 1980s, aimed at the introduction of more incentive-oriented salary policies, have yielded sparse results. And positive effects of broader reform programmes aimed at overhaul of civil services or general administrations are scarcely visible as yet. The short-term response of donors to the deepening incentives crisis—payment of local salaries or supplements—has allowed individual development projects to go forward, but at heavy cost in terms of exacerbated disorder in the incentive structure and in other ways.

Because a critical precondition for effective capacity building—a congenial work environment in the public sector—remains absent in most of Sub-Saharan Africa, donors and governments have to reconsider their strategies for the use of technical cooperation, and also their approaches to administrative reform. Failure to do so will mean continuing poor results, continuing disappointment with the slow pace of institutional development, and deepened African dependence.

As with all big problems, no simple solutions exist. We propose here a three-pronged strategy that should lead to an improvement of work environments and give capacity building a better chance to be more effective.

Confront the Salary Supplements Issue

Direct and indirect salary payments by donors have become a common and highly significant part of the administrative landscape in Sub-Saharan Africa. No technical cooperation strategy can ignore the problems and issues generated by this set of donor interventions.

Some preliminary observations should be made before turning to proposed policy changes. First, everybody agrees what the best solution would be: a comprehensive and quick civil service reform that would restore an acceptable level and structure of salaries. This would in principle remove the need for most donor supplements. But for reasons just reviewed, such a reform is highly improbable in the medium term in most countries. In addressing the supplements issue, as in all dimensions of technical cooperation for capacity building, the prudent assumption for policy making has to be that the problem of inadequate incentives for skilled and experienced public sector employees will not be resolved soon.

Second, there should be no blanket rejection of external financing of local salary costs on grounds of principle. Whether explicitly or not, most donor attitudes are based on what can be called the classic rationale for donor financing of technical assistance personnel, which parallels the rationale for capital assistance: that there is an absence (shortage) of relevant local skills and not enough foreign exchange to import them. Foreign aid then finances the offshore costs of technical assistance. The assistance creates human capital, strengthens institutions, and transfers technology, thereby generating faster growth. The higher level of output permits domestic absorption of any recurrent costs involved. This is the implicit model underlying the principle of "no local salaries" subscribed to by most donors.

But alternative rationales have come to the fore in recent years, and these provide good analytic support for donor payment of salary costs.

- One is the rationale of project cost minimization: donors should allocate aid resources optimally in implementing their programmes, without regard to whether inputs are local or offshore in origin. Local skills exist. Programme costs can be reduced by drawing on them.
- Another is the fiscal austerity rationale. Local skills exist, but governments cannot afford to pay for them. The underlying idea is illustrated by the Mali doctor story that circulates widely in West Africa: numerous (over 100) Malian doctors are said to be without employment at the same time that donors finance many expatriate doctors as technical assistance personnel, few of whom are specialists. On the surface, this is hardly cost-effective.
- The topping-up rationale is related. One important source of low productivity in implementing individual development projects, or in key recurrent activities such as tax administration, investment project choice, policy analysis, and primary education, can be addressed by salary supplements. The return on such expenditures is likely to be higher than in most other uses.

Other advantages accrue from use of nationals in technical cooperation projects—for example, greater ownership, deeper and broader learning by doing, and improvements in project performance because of greater knowledge of country-specific circumstances.

The arguments against salary payments are or should be grounded much less in principle than in experience; the negative empirical consequences are there for everybody to see.

- Salary payments quickly make a shambles of public sector salary structures. Ad hoc, uncoordinated, often unknown side payments and diverse forms of remuneration create a web of parallel wage structures covering a substantial percentage of the upper-level civil service. Project-financed staff make many times more than equally qualified staff on government terms. The grounds on which beneficiaries are selected are often not transparent, and many do not benefit. These systems are viewed by national staff as extremely unfair. Those who benefit are not always those who work hardest, and civil servants often say that pro forma payments are made to people at the top (Permanent Secretaries, Directors, and the like) even when they make little contribution to project-related work. The result: resentment spreads, grievances multiply and fester, commitment to the job declines, and productivity falls.
- One of the worst aspects of the current situation is that every donor for virtually every project is making its own rules, leading to destruction of salary and assignment procedures. Abuses are rampant. The 1985 Somalia study found individuals receiving supplements from several projects at the same time—not a rare practice if anecdotal evidence in other countries is to be believed. And the Somalia and Mozambique studies also confirmed the general belief that there is much competitive bidding by donors and projects for local staff with resulting higher job turnover.[22]
- In addition to this haphazard reshaping of salary structures, a supplement-ridden civil service makes it nearly impossible to build institutions and ensure sustainability and self-reliance. Nationals hired with foreign aid money have no secure roots in their organization; their aid support, here today, can be gone tomorrow when donor preferences change. For this and other reasons, externally financed hiring of nationals and payment of supplements induce churning in the internal labor market; many of the most dynamic people will tend to hop from project to project. Payment also upsets any government staffing

priorities as projects (and other donor employers) drain from the core civil service many of its brightest and best.[23]

- Payment of salaries and supplements can be regarded as contrary to the spirit if not the letter of ongoing adjustment programmes aimed at containing wage bills and slimming public sectors. This would seem to be the case when governments promise on the one hand to maintain a wage bill ceiling and maintain or reduce numbers employed, while on the other hand they hire staff with donor financing.

- Payment can also be contrary to related objectives of structural reform. For example, the 100-odd Malian doctors and others like them have to be persuaded that their future lies elsewhere than in the public sector. The chances are slender that government money will be available to hire them or sustain their work. Anything that encourages them to believe that this is not so, that a career awaits them in the public sector if they hold out long enough, slows that transformation of expectations that is fundamental to structural adjustment.

- External salary support can erode the will to reform. The patchwork system of wage supplements and imported technical assistance personnel allows government operations to continue at a minimum level of efficacy, so that pressures for deeper, more basic reforms are dissipated.

- Because nationals hired with external resources are not ensured of future employment when aid is withdrawn (the prospects for absorption by national budgets being so slight), sustainability of activities financed this way is thus unlikely.

- The relationship between expatriate technical assistance personnel and government counterparts, already under stress, is further strained when—as in many salary supplement arrangements—the expatriates become "paymasters."[24]

This formidable list of disadvantages explains why donors have been reluctant to support local salary costs. But dilemmas persist. In principle, some kinds of salary payments make good sense. And in practice, complete cessation of these payments is neither feasible nor unambiguously desirable. It would mean reduced development spending and reduced local participation in project implementation. It might also mean intensified withdrawal of effort among higher-level public sector employees as a result of lower average levels of remuneration. Because each donor has strong incentives to avoid these consequences, each would probably continue to provide supplements, probably in increasingly convoluted and disguised forms.

There are no easy ways to resolve these dilemmas. Most donors have dealt with them by establishing a two-track policy. As noted earlier, donor practice is to reject payment of local salaries in principle, but to allow freedom of action to projects they finance. UNDP, as the leading player in technical cooperation, has grappled more with this issue than others and has developed a more detailed and explicit set of guidelines than most other aid agencies.

UNDP policy allows special compensation to be paid to government personnel on projects for no more than six months, with payment for more than that period dependent on approval by the administrator, if the government agrees explicitly to take over this payment at a specific future date. This approach has several drawbacks. Governments often are unwilling to make commitments to increase the wage bill. Many times they are unable to do so because of competing macroeconomic policy objectives. And, even if governments were, higher-priority future needs may make it unwise or impossible for them to honor such commitments. In any case, this kind of conditionality is not credible because it is highly unlikely that noncompliance would be sanctioned.

What then should be done? Some guidance is provided by the 1991 DAC report on principles, which summarizes the consensus donor position on these issues.[25] After noting that low salaries induce inefficient civil service operations, the DAC report says that donors should not top-up salaries but encourage governments to address civil service reform. The report also notes that coordinated multidonor action is especially desirable in this area. Its main policy prescription is as follows:[26]

> As a matter of principle, the practice of salary supplements should be avoided. It exacerbates wage distortions and intensifies donor

competition for scarce recipient administrative skills. In exceptional cases where the recipient government and donors explicitly agree that they are essential, salary supplements or fringe benefits provided for similar purposes should only be considered, provided that they are clearly time-bound, that they follow explicit rules, are fully transparent, and that donor practices are harmonized, possibly through a comprehensive and centrally administered mechanism to which [sic] both the government and the donor community agree. Reform measures should be made part of mutual commitments that could be monitored regularly. Exceptions should be seen as temporary measures pending completion of a civil service reform process.

Although this statement provides some helpful guidance, it leaves the main policy dilemmas in place. Donors still have to assess the advantages and disadvantages of local salary payments, and determine whether and how traditional objections to this form of aid should be overridden. The report recommends that donors not pay local salaries unless recipient governments adopt civil service reforms. But such process conditionality has proved difficult to monitor in a meaningful way. And in any case civil service reform is hardly a one-shot or short-term event. Moreover, adoption of these reforms is no guarantee of implementation. Experience indicates high potential for failure and, in the best circumstances, a long, slow process of improvement.

Presumably donors would continue to provide supplements, according to the recommendation of the DAC report, as long as the administrative reforms are judged to be on track. But this opens a Pandora's box of questions, and many objections. Is it realistic, for example, to assume that these kinds of institutional reforms can be objectively monitored or that all donors would ever agree to pull out because of nonperformance? (This happens rarely, only when there has been massive, blatant, and unrepentant violation of conditionalities.)

More specific, and perhaps more workable guidelines can be set down in the general framework of DAC policy recommendations. Most important, clear distinctions have to be made between those types of activities for which donor financing of salaries is justifiable and those for which it is not. Payment of nationals for work on projects is clearly justifiable, for example. Here the benefits seem to outweigh costs and risks. To reduce costs and to increase the efficacy of training and technology transfer, wide use of nationals is desirable as project staff and as consultants.[27] One major disadvantage persists: a dual salary structure is created. But this is unavoidable. If the numbers involved are relatively small, the impact can be contained.

External financing of nationals for project work is desirable only if "projects" are correctly defined—as activities with a beginning and an end. Some flexibility is possible; start-up periods for completed projects might be included. But the conventional rule on the undesirability of external financing of ongoing, operational costs applies with full force to the salaries issue.

That some local costs are commonly financed as part of technical cooperation projects does not change the argument. Nor does this position contradict the earlier proposal, in Chapter Five, to give increased attention to strengthening transport, secretarial, materials supply, and other internal support services. This assistance would be time-bound, in the framework of administrative reform projects.

The issue becomes fuzzier when support to whole sectors or subsectors is concerned. What can be said, for example, about the payment by some donors of recurrent salary costs of the primary education system (as in Tanzania, for example), on the grounds that the system is so starved for resources that it is nearly nonfunctional, and this puts at risk the country's future development.

The spirit of the DAC recommendation is that this kind of assistance has to be conditioned, as in World Bank SECALs. There are risks even in this position. Much lending for the education sector has gone on in the past decade, usually with conditionality aimed at shifting resources from the university level to the primary level. But these conditionalities are often not implemented. And few African countries have yet to reallocate in that direction, despite a strong intellectual consensus that it would be efficient and equitable to do so. The hard-headed question has to be asked: does external financing of local salaries allow policy makers to put off politically difficult but economically and socially desirable decisions of this type? Put another way, why shouldn't governments themselves move to deal with problem sectors so critical to social and economic development as primary schooling?

What about salary supplements as distinct from hiring of nationals? Various proposals for change have been put forward. One recommended policy change is to allow payments for a restricted set of countries (the least developed), and only if they are under clear fiscal duress, if other implementing vehicles have been looked at and if project start-up or effectiveness would be seriously hampered. Only in-kind payments would be permitted (housing, training, and transport, for example), except in extraordinary circumstances. The payments would have to be for extra work, would be specified in project agreements, and would in no case continue to be paid by UNDP beyond the life of the project.

One serious problem with this proposal is that in many cases the donor is asked to pay supplements to allow government project staff to perform their normal duties. These in-kind arrangements also suffer from lack of transparency.

The most widely discussed idea is the creation of an extrabudgetary facility that would centralize all payments of salary supplements to government employees. Donor resources would be pooled in a common fund and donor coordination achieved via participation in the running of this facility. This would be a stopgap measure, tied to civil service and general administrative reforms. It or something like it is implicit in the 1991 DAC report guideline cited above.

This fund might be patterned on the Bolivian model. The donor community in that country financed a $6-million foundation to pay salaries of 500 high-level government posts. This financing was extrabudgetary and not subject to IMF ceilings on the wage bill. This kind of arrangement is consistent with ideas being put forward for possible solutions to the salary supplements problem in African countries; it is in line, for example, with the rather cryptic DAC suggestion given above. Unfortunately, information is not at hand to look at this innovation in detail, or to determine how well it is working.[28]

In Uganda, the desirability and practicality of the common fund concept have been the subject of much debate. Donors in that country objected to the proposal to set up such a fund on the grounds that it required too much detailed control over their projects, and, like a cartel, all would have to agree to the rules. The required controls over performance would involve unacceptable and undesirable donor intrusion. In any case, disguised competition would replace direct supplements.

The Uganda case suggests also that the common fund idea may not be acceptable to governments, and could be counterproductive. It might be a diversion from the main task of basic civil service reform and a source of diminished commitment to this reform. This was the view of the Uganda Public Service Review and Reorganization Commission, which criticized a 1990 common fund proposal on these grounds.[29]

The evidence to date suggests that even were donors to work out more coordinated and rational schemes for paying salary supplements, they would continue to have strong negative impact on the operation of civil services. This is suggested by the experience of the Ghana Skills Mobilization Scheme, a World Bank-financed project, which—as part of the country's adjustment program—aimed at increased use of local skills. The roadblocks that emerged show how difficult it is to win acceptance even of well-conceived salary supplement arrangements.

- The scheme provided for recruitment of local consultants for specific tasks. A roster of consultants (41 as of late 1989) was established and standard rates set depending on qualifications. Serving government employees could not be used. A few studies were commissioned under this component, but the number of consultants used was small—three, as of late 1989.[30]
- Ghanaians from the private and parastatal sectors could be recruited for long-term employment in central government and paid their previous salary plus a premium equivalent to $200-$300 a month. This provision generated so much discontent in the civil service that it was never implemented.

- A relocation allowance ($10,000), transport costs, and a salary premium were made available to induce overseas Ghanaians to take public sector jobs at home. As of early 1991, only one person had been attracted—and no salary premiums were paid.
- Special duty allowances were given to serving civil servants for special tasks related to the Economic Recovery Program. Between 1987 and 1989, 90 to 150 employees a year received between $25-$150 a month in topping-off. Resentment of nonrecipients was again a serious problem, which put this component in doubt.

All of this suggests there is good reason for the hesitancy and the deep reluctance of donors to ratify the payment of salary supplements as official policy. They are profoundly unsettling, even when carefully conceived. More orderly and transparent approaches like the Bolivian extrabudgetary foundation are as yet untested and, where proposed in Africa, have been resisted by many donors and, in the case of Uganda, by the commission overseeing civil service reform.

The best policy on salary supplements, then, is damage control. Supplements should be contained and made more transparent. Improvement can come from greater knowledge and from informal donor coordination. The multiplication of country studies on the nature and extent of these payments would help.

The best path to a satisfactory enabling environment for technical cooperation is for donors and governments to focus on the basic problem of creating stronger institutions to meet development needs. They should pursue, as quickly and as vigorously as possible, a medium-term reform strategy with two main elements: faster and better rehabilitation and reform of civil services and, at the same time, the acceleration of steps to downsize the public sector.

Intensify Efforts to Reform Civil Services

Formidable difficulties confront civil service reformers in Sub-Saharan Africa, perhaps more than elsewhere, given the extent of deterioration in many countries, the sparsity of resources, the region's soft politics, and the tremendous obstacles inherent in the task. This is hardly headline news. Nor is it particularly original to conclude that only one policy implication can be drawn from it: reform efforts have to be improved and intensified. The present situation in much of the continent is plainly unacceptable. It presents intolerable obstacles to the emergence of sustainable capacity in management and policy. To allow it to persist is to condemn much of the region to poor governance, slow growth, and deepened dependence. It incidentally also condemns capacity-building efforts to continuing ineffectiveness.

Everybody agrees that reformers can do better. They can devise more appropriate changes. They can invent more refined tactics for implementation. They can draw more effectively on lessons of experience, from both successes and failures.

We cannot enter into details here for many reasons, not least the fact that the scope of the problem is too vast. At issue, after all, is not only wholesale administrative reform but also a mass of special areas of public sector management: for example, how to organize finance ministries, planning agencies, and budget bureaus; how to organize investment project selection and programming; how to reorganize transport pools or debt management systems or management information arrangements in general; and how to devise new frameworks for government-state enterprise relationships.[31]

A few general points, however, merit attention. These are put forward here in full awareness that they are incomplete and that they represent only one set of priorities among many others that are possible.

- Civil service reform should itself be a principal focus of technical cooperation efforts in coming years. It is already a priority area of technical cooperation activity, but should be much expanded.
- According to some observers, recent reform efforts have been too dominated by short-term preoccupations, notably job retrenchment. It is certainly important to slim down the public sector. But the costs are high: on-the-job malaise, spread of contentiousness to the overall reform programme, and diminished political enthusiasm for the reforms in general. The benefits seem small. The age distribution of many African civil

services is such that attrition by retirement may make the problem more manageable in the next decade. And there may be more effective indirect ways to proceed, as is indicated in the downsizing discussion below.
- Related to this, reform of personnel management has probably received a disproportionate share of donor attention. We argued earlier that major efforts should be devoted to improving supply management and support services: computerization, photocopying, secretarial services, motor pools, repair and maintenance, and the like. And other key areas seem to be receiving too little attention: planning and budgeting; and investment project implementation, including procurement systems.
- The prospects for success in administrative reform would be greater if it were distanced from externally imposed conditionalities. Without local ownership, the best administrative reforms are easily undone and conditionality undermines the growth of local ownership.[32] Besides, the "process" type of conditionality that is typical in administrative reform programmes is rarely effective because assessment of performance is too subjective. And to the extent that conditionality makes the dialogue contentious, when it should be a joint search for solutions, conditionality is counterproductive.

General administrative reform and even comprehensive civil service reform face so many pitfalls that alternative tactics and strategies based on more partial approaches should receive greater attention. One approach is to undertake piecemeal reform, focusing on a few critical functions or agencies and creating reform islands or enclaves.

The strategy here would be to isolate for special treatment bureaus or agencies whose effective operation has particularly far-reaching spillover effects on the overall operations of the public sector. The staff in these agencies would then be given much better conditions of work, including salaries above civil service averages. Examples of possible high-wage enclaves or islands might be customs administration, other (inland) tax collection agencies, project evaluation bureaus in planning ministries, budget agencies and other key finance ministry bureaus, policy analysis units, and medical services.

Early results from experiments involving creation of such reform islands seem encouraging. The most dramatic example is that of the Ghana National Revenue Service. Organizational changes, including improved remuneration and job conditions, have led to extraordinary improvements in efficiency in that agency. Box 6.1 gives details.

BOX 6.1

REFORM ENCLAVES: GHANA'S NATIONAL
REVENUE SERVICE

> Between 1984 and 1988, major organizational changes were introduced in this agency, which is responsible for all of Ghana's tax collection. Aggregate revenue increased during this period from 6.6 percent of GDP to 12.3 percent. All tax categories showed large gains. Efficiency (apparently measured by collections per person or per dollar of budget) rose by 30 to 40 percent annually over these four years.
>
> This achievement is said to be explained by the following factors.
> - Good management—an outstanding leader at the top;
> - Management autonomy, notably in salary determination and in the right to use a share of tax collections for agency spending;
> - A better incentive framework for employees. Base salaries were comparable to those in the civil service, but allowances (in cash and in kind) were much higher. External training opportunities were made available in abundance, as an incentive device;
> - Because of its budgetary autonomy, the National Revenue Service could offer much better working conditions than available elsewhere—supplies, office space, and the like;
> - Early in the programme, the agency fired many excess workers and recruited better-skilled and better-motivated workers; and
> - The agency developed effective information-processing technologies.
>
> Source: Louis de Merode, *Civil Service Pay and Employment Reform in Africa: Selected Implementation Experiences*, World Bank, 1991.

The enclave approach has important advantages over other partial reform tactics, such as the setting up of a common fund for salary payments to critical personnel or the type of across-the-board effort represented by the Ghana Skills Mobilization Scheme. The enclave approach avoids or much dilutes the problem that cripples individual merit pay schemes in public sectors: the difficulty of agreeing on who is more productive and by how much. It also reduces the jealousies and resentments created by the scattered presence of favored employees. It involves relative physical separation of beneficiary groups—everyone in the office benefits, so to speak. And it combines salary increases with other administrative reforms for the unit in question.

Even with the adoption of these or other possible proposals for improvement in the effectiveness of administrative reform, the enterprise will remain high risk and long term. It is essential, therefore, that back-up strategies be put in place. In the short run, for example, it makes little sense to count on stable relationships with counterparts when slack commitment and high job turnover are likely to persist. And, more fundamental, technical cooperation providers and African governments have to do their planning and their technical cooperation programme designs on the assumption that over the medium term—10 to 15 years—public sectors are almost sure to remain in fiscal duress and work incentives are highly likely to remain inadequate.

Because substantial improvements in the public sector work environments of much of Sub-Saharan Africa cannot be expected in the medium term, donors should follow a parallel track even while financing major efforts at administrative reform. While these efforts are under way, before they yield substantial results, donors have to help governments find alternative ways to increase and improve the provision of key public services.

These alternatives are at hand. Some have been tried in the experimental reform programmes of recent years. Others have been extensively discussed and await more systematic application. The basic idea is to reduce public sector burdens by encouraging other agents to do more—profit-seeking private individuals and organizations, semi-public bodies, or nonprofit (voluntary) agencies. The instruments are well known—especially deregulation, contracting-out, and internal divestiture of activities not central to the mission of government.

Downsize the Public Sector

Donors and governments should recognize that technical cooperation for capacity building does not mean that the capacity has to be in the public sector. Indeed, for many activities it may be possible to build sustainable capacity only outside the public sector. This is not a question of ideology. It is common sense and springs from recent experience. Everybody has learned that creating effective organizational capacity in the public sector is far easier than keeping that capacity alive when outside aid ends. In fiscally crippled public sectors, operational effectiveness depends too often on salary supplements and other budget support provided by donor projects.

With outside aid, for example, governments can create an organizational capacity for rural well digging and even maintenance. But given the low likelihood of effective performance of such an organization when aid is terminated, the technical cooperation invested in building public sector capacity of that kind is likely to be wasted. A technical cooperation programme that nurtures private rural artisans or plumbers with training, credit, and financing of initial contracts with villagers is much more likely to leave something behind. The situation is the same for many other activities. A data processing capacity can be created in a finance ministry, but it will have difficulty surviving without a reliable maintenance budget and in the absence of decent salaries for staff. A private or semipublic data processing entity stands a far better chance.

Many examples of successful downsizing through privatization, contracting-out, and the like are cited in casual discussion, although they are mostly unstudied and undocumented. One documented case exists in Gambia, however.[33] "Management reviews" of government agencies in that country looked at their operations to identify overlapping, obsolete, or low-priority functions. The reviewers also looked at alternative forms of goods production and service delivery—in-house production, management contracting, contracting-out, and privatization via divestiture, for example.

A review of Gambia's Ministry of Agriculture resulted in a reorganization of the ministry around its core functions, the transfer of lower-priority activities to other agencies, and the privatization or contracting-out of several important activities that could be privatized: seed multiplication, fuel supply, building maintenance, vehicle maintenance, some veterinary supply services, crop spraying, pig and poultry production, tractor hire services, and veterinary services. The ministry's staff fell by a third as a result.

A similar concentration of energies and slimming down took place in the Ministry of Public Works. A carpentry shop producing office furniture was privatized, registries and other administrative functions were consolidated, and central storage of materials was dropped. These and other changes led to a reduction of 25 percent in the ministry's staffing.

These approaches have great potential, still largely unexploited, to reduce public sector burdens and to increase the supply of public services and the efficiency with which they are provided.

- The range of possibilities for contracting-out is great—data processing, maintenance of all kinds, subactivities in ports (warehousing, stevedoring, and rat control), and research and policy analysis, to name just a few.
- Internal divestiture, or privatization by fragmentation, provides an equally vast set of opportunities for unburdening the state. As the Gambia story suggests, activities exist in every ministry or state-owned enterprise that are peripheral to the central functions of the agency or organization and that can be liquidated or privatized. Virtually all commercial activities are possibilities, and many services—cattle farms; slaughterhouses; poultry farms; repair shops in national railway or port authorities; railway, bus, and machinery maintenance; rural water supply; and well maintenance. In many cases, existing groups of public sector employees could be encouraged to form private companies or cooperatives to take on this work, with initial support by external donors.
- Durable capacity in critical areas can be created by encouraging the formation of private or semiprivate foundations or institutes that can pay decent salaries and build the right environment for effective performance—and for effective

technical cooperation as well. One model often mentioned is the Vargas Foundation in Brazil. Unable to pay adequate salaries and provide attractive ancillary conditions necessary to recruit and hold competent policy analysts, the Brazilian government formed this foundation to house economists who do policy studies and other research, largely on contract with government and the private sector. Many such autonomous institutes are under consideration in Africa.

- More general and rapid deregulation, particularly in the social sectors, trade, and transport, would lighten the management burden on the public sector and increase the supply of services. Restrictions on private provision of education and health services should be reduced. The public-private mix in urban transport and in export and food marketing should be reconsidered, with a view to accelerating the move to private suppliers.
- Efforts to encourage formation and strengthening of local consulting firms should be expanded. The urgency of these efforts is widely acknowledged. But progress in many countries seems slow or token. Donor procurement rules and practices are crucial here. Nurturing mechanisms are straightforward and could be strongly expanded—for example, through more numerous, smaller contract awards and twinning arrangements with foreign consulting firms.
- Donors should give greater attention to strengthening centers of excellence that can provide training research and other services on a regional basis. When established national institutions exist, their regional mandate should be expanded. For example, Senegal's engineering school already enrolls many non-Senegalese and could take more; facilities for engineering training in neighboring countries could be tailored accordingly. In the allocation of responsibilities for agricultural research, regional and international research centers could take greater responsibility for activities normally left to national centers in cases where their effectiveness is being undermined by low salaries and low, unreliable budgets.

The need to slim down the public sector to improve efficiency and increase the prospects for successful rehabilitation of civil services thus has a regional dimension. The provision of public goods could be much more actively allocated on a regional basis, with many benefits. Specialization and economies of scale would lead to better, cheaper services. The movement toward closer regional integration, a high priority for governments and donors alike in the 1990s, would be stimulated.

The movement toward greater use of nongovernmental organizations should be encouraged. True cooperatives, self-help associations, economic, social, and political interest groups of various kinds are sources of economic initiative and political liberalization. Their growth will help develop a diversified civil society, multiple centers of initiative and economy activity, and a diversity of opinion and analysis central to democratic government and better decision making in the public sector.

The benefits from these diverse approaches to a downsizing of the public sector would not be only, or even mainly, a reduction in public wage bills; average and total labor costs financed via the budget may increase in the short run. But entrepreneurship will be encouraged and specialization will grow. Greater dynamism and responsiveness in supply and greater competence and efficiency in provision of services should result. In the longer run, average costs should fall. And the smaller public sector, with fewer civil servants, should be more amenable to rehabilitation and to the introduction of appropriate salary incentives.

Two benefits from downsizing the state are particularly pertinent from the perspective of donors anxious to help build durable capacity. Privately based organizational competence is not likely to disappear as foreign assistance diminishes. And the task of civil service reform (and the strengthening of public sector management in general) would be simplified and made more manageable, hence more likely to succeed.

NOTES

1. The examples come from the summary of technical assistance proposals for 1991-1994 in Swaziland "Development Plan 1991/1992-1993/1994," Economic Planning Office, Mbabane, February 1991.

2. The magnitude of these improvements is usually small; it consists of the new knowledge incorporated in reports and the differential efficacy of technical assistance personnel compared with local personnel. Impacts, moreover, are usually ephemeral: reports gather dust and their information becomes obsolescent.

3. United Kingdom, Overseas Development Administration, "Civil Service Reform in Sub-Saharan Africa," record of a Conference Organized by the Overseas Development Administration, March 29-31, 1989.

4. See Derek Robinson, *Civil Service Pay in Africa*, ILO, 1990, Chapter Seven, for a discussion of this point.

5. Uganda is perhaps the most dramatic case. In 1990, the lowest established civil servant rate was $6 a month at the official exchange rate; the salary of the highest civil servant—say a permanent secretary—was $32 a month. A public sector review commission report in 1990 concluded that civil servant's wages were about 1 percent of a living wage. Clay Wescott, S. Haregot and I. Coker, "Rationalization of Financial Support to the Civil Service of Uganda," UNDP, May 1990, p. 4.

6. The data in Table 6.2, which are taken from a 1990 ILO study, indicate wider differentials than those given in various World Bank documents, notably the study by Louis de Merode, "Civil Service Pay and Employment Reform in Africa: Selected Implementation Experiences," Division Study Paper no. 2, Institutional Development and Management Division, World Bank, June 1991. The differences in the size of the salary differentials seems to be due to dissimilar definitions of high and low salary receivers.

7. Barbara Nunberg and John Nellis, with L. de Merode, "Civil Service Reform and the World Bank," CECPS, World Bank, December 1989.

8. Deteriorating wage incentives have also induced African governments to find ways to escape civil service salary constraints. They have introduced nonsalary and noncash benefits such as privileged and

subsidized access to key goods (basic food staples and gasoline, for example), fellowships abroad, and trips that bring high per diems. They also expand the number of contractual (nonestablished) employees who benefit from more generous salaries. Governments, through the recipient technical agencies, also put pressure on donors to finance local operation costs and hence improve material conditions of work. In some cases, they push for salary supplements. Many governments have recently begun to loosen civil service regulations regarding employment and contracts, allowing employees to take outside assignments.

9. When salary supplements are introduced, donors try to limit their duration. UNDP, for example, allows special salary supplements to be paid to civil servants only in the formally designated "least developed countries," for no more than six months and only with special approval from the administrator.

10. Donors define supplements as "direct cash payments made by donor organizations to a national employed by a national organization in a Third World country, but involved in implementation of a project or program financed by the donor."

11. Nunberg et al., 1989; and de Merode, 1991. Most were part of structural adjustment loans; the remainder were free-standing technical assistance loans.

12. See, for example, Nunberg et al., 1989; de Merode, 1991; and Trevor Robinson, "Ghana Civil Service Reform Programme," in U.K./ODA, "Civil Service Reform . . . ," 1989.

13. de Merode, 1991, p. 25.

14. Nunberg et al., 1989, Table 8.

15. de Merode, 1991.

16. The U.K. public administration program in East Africa has worked in Uganda since 1983, helping four main institutions: the Ministries of Finance, Public Service, Planning, and Local Government. The program supports also the Uganda Institute of Public Administration and the Auditor-General's office. Over the eight years, 1983-1990, the cost of these activities (including related training in the United Kingdom) was about $7 million.

17. Nunberg et al., 1989.

18. World Bank, Operations Evaluation Department, "Free Standing Technical Assistance for Institutional Development in Sub-Saharan Africa," Report no. 8573, April 1990.

19. World Bank, "Managing Technical Assistance in the 1990's. Report of the Technical Assistance Review Task Force," 1991, Annex 6.

20. See, for example, Gelase Mutahaba, *Reforming Public Administration for Development: Experiences from Eastern Africa*, African Association for Public Administration and Management, 1989, Kumarian Press, Hartford, Conn., pp. 34-39.

21. See the following samples from a vast literature. Gerald Caiden, *Administrative Reform*, Chicago: Aldine, 1969; by the same author, "Administrative Reform: A Prospectus," in *International Review of Administrative Sciences*, 44:1-2, May 1978; J. Montgomery, *Sources of Administrative Reform: Problems of Power, Purpose and Politics*, Bloomington, Indiana, 1967; A. Adediji, "Strategy and Tactics for Administrative Reform in Africa," in A. Ryweyemamu and G. Hyden, eds., *A Decade of Public Administration in Africa*, Nairobi, East Africa Literature Bureau, 1975; and Y. Dror, "Strategies for Administrative Reform," in A.F. Leemans, ed., *The Management of Change in Government*, The Hague, 1976.

22. UNDP and World Bank, "Report of the Technical Cooperation Assessment Mission: Somalia," 1985. Similar findings have been reported in studies in Uganda ("Discussion Paper on Technical Cooperation," prepared for Uganda Consultative Group Meeting, 1990) and in Mozambique (M. Rajana, "Report on Remuneration and Conditions of Service Offered by Expatriate Agencies Operating in Mozambique," UNDP, March 1991).

23. This was borne out in the 1985 UNDP and World Bank study of the impact of technical cooperation in Somalia, which found that in the vital rural development sector the central ministry was stripped of good staff because it was the "poor cousin" of agricultural institutions; all the money and the good jobs were in the rural development projects.

24. A recent study of salary supplements and related issues in Uganda states: "The incentive program is placing donors increasingly in the role of supervising and controlling the daily work and lives of civil servants, creating an atmosphere not unlike the pre-independence era." Wescott et al., 1990, p. 40.

25. OECD, DAC, "Principles for New Orientations in Technical Cooperation," Paris, 1991.

26. Ibid., par. 26.

27. However, it is not a good idea to finance the salaries of nationals who work as de facto counterparts—a common practice.

28. Nunberg et al., 1989, p. 20.

29. Wescott et al., 1990, p. 39.

30. The information comes from de Merode, 1991.

31. See Mary Shirley and John Nellis, "Public Sector Management Reform, Lessons of Experience," World Bank, 1990, for a brief overview.

32. This makes especially worrisome the observation in a recent World Bank evaluation of civil service reforms in three West African countries that, without conditionality, the governments concerned would not have cut numbers in the civil service or raised salaries of higher ranks as much as they did. See de Merode, 1991.

33. The case is described in de Merode, 1991, p. 37.

7

SUMMARY AND CONCLUSIONS

Two generations ago, technical cooperation was universally esteemed. Early students of economic development and all who were concerned with world poverty believed it to be a straightforward and powerful device—one that could and would make major contributions to the economic transformation of developing countries. President Harry Truman's "Point Four" proposal, put forward in 1949, reflected this early vision. At the very least, technical assistance or technical cooperation was recognized as the indispensable handmaiden to capital investment. Technical cooperation was needed not only to help build roads and universities but also to help develop the local capacity to maintain or run them.

Questions about the efficacy of technical cooperation were not entirely absent in the ensuing decades, but these were few and gentle. In the press, in the literature on development, in the cabinet meetings of African governments, in the corridors of aid agencies, and in the legislatures of the industrialized world, technical cooperation was seen as a good thing, a relatively well-functioning mechanism that met essential needs for training and technology transfer.

THE ATTACK ON TECHNICAL COOPERATION

Quite suddenly, all this has changed. In the 1980s, technical cooperation has been subjected to a growing barrage of criticism. Numerous reports and evaluations, coming from recipient countries and donors alike, confirm the existence of generalized discontent with the performance of technical cooperation. Dissatisfaction is greatest in heavily aided countries with low incomes, especially in Sub-Saharan Africa.

Many of the problems and shortcomings of technical cooperation have become more evident as a result of more and better diagnostic studies at the country level—notably as a result of analysis sponsored by UNDP under NaTCAP. Internal assessments and evaluations of the main multilateral aid agencies and many bilateral donors have also revealed deep concern over the magnitude, form, and effectiveness of technical cooperation, particularly in Africa. Consensus views of donors on the need for change are synthesized in the OECD/DAC 1991 report, "Principles for New Orientations in Technical Cooperation." This report summarizes the reasons for recent dissatisfaction and suggests new directions for the future.

The criticisms of technical cooperation cover a wide range, and donors and recipients emphasize different aspects. But there is much overlap. Almost everybody acknowledges the ineffectiveness of technical cooperation in what is or should be its major objective: achievement of greater self-reliance in the recipient countries by building institutions and strengthening local capacities in national economic management. Despite 30 years of a heavy technical assistance presence and much training, local institutions remain weak and this type of assistance persists. Deficiencies in technical cooperation are not the sole or even the main explanation for this situation, but they contributed significantly to it.

This conclusion, which is suggested by the observation and common experience of practitioners and local participants, is a major theme in NaTCAP-sponsored diagnostic studies and is borne out by donor evaluations. A recent worldwide review by the World Bank, for example, found that only one in four institutional-development components of completed technical assistance projects could be judged "satisfactory," and 42 percent had negligible impact. Results in low income countries are worse: of 369 African projects with institutional development components, only 22 percent had "substantial" results.

A second criticism is that technical assistance is very costly, at both the macroeconomic and the project levels.

- In Sub-Saharan Africa, technical cooperation grants amounted to $3.2 billion in 1989, or a quarter of total official development assistance in the region. In some countries, the proportion is much higher. And this is an underestimate because much technical assistance is excluded—for example, loan-financed assistance like that provided by the World Bank, some technical assistance provided in the framework of capital projects, and non-OECD assistance. The true total was probably well over $4 billion annually at the beginning of the 1990s. In several countries, amounts spent on technical assistance personnel substantially exceed the total wage bill of the national civil service.
- Resident expatriate personnel have become extremely expensive. In many countries of the region, the cost of a resident expatriate adviser now approaches $300,000 a year.

Political leaders, officials, and intellectuals in most African countries are increasingly restive about the persistence of technical assistance in its present magnitude and forms. Many find it intolerable that local university graduates should be jobless while foreign technical assistance absorbs large sums. They feel that many technical assistance personnel are imposed by donors, that the vastly higher salaries paid to these personnel are rarely justified by productivity differences, and that it is unsettling when any foreign assistant of modest rank receives a salary 10 times higher than that of the most senior cabinet minister. They, and many donor representatives, see technical assistance as a disguised and inefficient subsidy to operating budgets; it allows hard-pressed governments to employ foreign staff in place of available trained nationals who cannot be hired because of wage and employment freezes and similar measures reflecting severe budget constraints.

In its present form and magnitude, technical cooperation is often institutionally unsettling.

- Its donor-driven character contributes to disorderly decision making within African public sectors; ministers of finance and planning have difficulty implementing nationally decided priorities when so many key decisions on critical personnel are made in donor capitals.
- It finances activities that are not sustainable.
- It contributes to the breakdown in civil service salary structures and the growth of job instability that has resulted from donor payment of salary supplements and hiring of nationals.
- The coexistence of well-paid, usually highly motivated resident expatriates who have access to vehicles, office equipment, and supplies as well as much easier access to information, with poorly paid, underequipped, and hence frequently unmotivated national civil servants often deepens dependence on foreign experts and relieves local staff of responsibility.

SOURCES OF INEFFECTIVENESS

Although donor and recipient country officials and observers may emphasize different aspects of the problem, there is a wide area of agreement about the specific sources of technical assistance failures. The consensus diagnosis focuses on four main problem areas:

- Weaknesses in design, implementation, and supervision of technical cooperation projects;
- Excessive reliance on one model of delivery for technical assistance—the resident expatriate-counterpart model, which has failed as an instrument for capacity building;
- The donor- or supply-driven nature of technical cooperation, which has led to excessive use, inefficient allocation, weak local ownership, and hence limited commitment; and
- Poor incentives and working conditions in recipient country public sectors, which lead to low local staff job motivation and high turnover, creating a Teflon-like work environment in

which capacity-building and institutional-development efforts fail to take hold.

CONSENSUS REFORM PROPOSALS

From this widespread accord on the definition of basic problems, a common set of proposals for change has emerged—not quite a standard package, but close to it.

1. Deliver the Existing Package More Effectively

The central idea here is straightforward: donors are urged to address nuts-and-bolts types of deficiencies that are pinpointed in virtually every study of technical assistance done in the past two decades and admitted by most practitioners. The proposed changes indicate the nature of these deficiencies.

With respect to project design, donors should give more time to preparation and give the training component more priority, seek greater local participation early in the design stage, take account of local institutional needs and capacities, avoid overambitious targets and automatic resort to top-of-the-line technology, avoid too-loose terms of reference, specify measurable outputs in these terms of reference, and emphasize capacity-building objectives more than operational support.

Implementation has to be improved by faster, better recruitment; giving greater weight in recruiting to qualities other than technical competence—cultural sensitivity and interpersonal skills, for example; working harder to ensure a counterpart presence; giving closer and more persistent attention to implementing training components; and supervising and evaluating better.

2. Change the Mix of Delivery Modes

The basic vehicle for on-the-job training and technology transfer has been the resident expatriate expert associated with a local counterpart. This model has proved to be deeply, irreparably flawed.

The experts are recruited mainly on the basis of technical competence; they are often good at their job but bad as trainers or on-the-job coaches. Many incentives drive them to give priority to getting the job done, while inducements to train and build local capacity are few.

In addition, the expatriate experts are older and better paid, and they usually have more training and experience than their local counterparts. Counterparts are almost always very badly paid and hence are forced to search for other ways to make ends meet. Even when they are committed to the work, counterparts have few incentives to stay on the job because their careers are best served by changing jobs. Expatriate advisers are also plugged into the ubiquitous expatriate information networks, so they often know who is saying, doing, or writing what and have easier access to local decision makers. They control project money and equipment. It's not surprising that local managers often turn to technical assistants for operational help.

Powerful forces thus pull technical assistance personnel in the direction of producing non-capacity-building outputs: the fact that these outputs are what the experts' bosses want and that the experts' performance in producing these "hard" outputs determines how they are evaluated; the job orientation and internal values of the experts themselves; and the frequently uncongenial environment for capacity-building efforts—low job commitment of counterparts, instability of staff and leadership, and lack of budget support.

All of this makes the resident expert-counterpart model a highly dubious instrument for capacity building. And it has other, even more fundamental deficiencies. In standard on-the-job training schemes, the trainee is given operational tasks and responsibilities under the supervision of a more experienced professional, who guides and checks performance. But this is not the way the expatriate expert-local counterpart arrangement works. Counterparts can rarely learn by doing. They are overshadowed by the expert. There is also hierarchical confusion. In theory, the expert advises his counterpart. In practice, the expert does much of the work. In most training systems, hierarchical lines between trainer and trainee are much clearer. It is perhaps not surprising that the resident expert-counterpart training arrangement is found nowhere except in the world of technical cooperation.

The policy conclusion that follows from this diagnosis is that reliance on the resident expert-counterpart model should be drastically reduced, if not abandoned entirely. Three substitutes are commonly proposed.

- Greater use should be made of short-term advisers and coaching arrangements. The pure coaching model involves intermittent short visits by expatriate experts working with national staff on specific assignments. During an initial mission, the expatriate expert helps define the problem and, with local staff, begins to attack it. He or she then leaves, remains in electronic contact with the project, and returns every few months to review progress with responsible national staff. The modified coaching model involves one resident expatriate who recruits and helps coordinate the work teams of short-term advisers.
- Much greater use can be made of local consultants. Consulting competence exists in most countries of the region. The main obstacles to its more intensive use lie on the donor side: unfavorable contracting practices and regulations, and inertia.
- Institutional twinning should become a much more favored instrument. The pairing of organizations with similar operational functions has great appeal. It gives assisted institutions continuing and systematic access to people and know-how accumulated in partner institutions.

3. Strengthen Local Management of Technical Cooperation

In almost all African countries—and in low income countries elsewhere—aid donors orchestrate the technical cooperation show. They conceive most project ideas, arrange their design, hire most of the experts, and oversee implementation. They also supervise and evaluate the project. This is the meaning of the frequently voiced criticism that technical cooperation is supply driven.

This situation has many costs and inconveniences. The most general and significant is that African authorities feel little ownership of activities with which they have been so little involved, making commitment problematic. But there are other costs in reduced effectiveness of public sector decision making and economic management. Governments are often imperfectly informed about their technical cooperation involvements. The programming and budgeting for this form of aid are much less developed than they are for capital projects, despite the sizable share of technical cooperation in total aid flows. Because screening systems for technical cooperation are extremely weak or nonexistent and there are usually many hundreds of discrete projects, the priority-setting agencies (finance and planning ministries) are unable to avoid acceptance of low-priority, overlapping, poorly conceived projects.

Most African countries receive development assistance, including technical cooperation, from so many donors that they are unable to manage effectively the resources at their disposal. One result is to encourage continued donor control and management of their own programmes, which means more control-type of technical cooperation and less capacity-building effort. Each donor has its own accounting and reporting requirements; compliance absorbs much of the available planning capacities.

Almost everybody agrees about the basic remedy for these problems: much greater responsibility for management of technical cooperation has to be transferred to local hands. Three recommended approaches are commonly put forward by participants and observers.

Voluntary Donor Transfer of Managerial Authority

Several bilateral donors are trying to hand over responsibility for management to recipient governments, and UNDP has long called for "national execution" of aid projects. Belgian aid authorities have experimented with co-management, under which they withdraw entirely from specific aspects of project management. They undertake only ex post control and some responsibility for financial management. The Nordics have called for conversion of the donor role, so that donors become mainly financing agencies with analytic and evaluation capacities. Host governments would take over management of all phases of the project cycle.

Stronger Local Management Via NaTCAP

The UNDP-sponsored NaTCAP process has been adopted by 34 African countries as of mid-1992. Its objectives are to introduce the procedures and build the organizational capacities that recipient governments need if they are to manage technical cooperation more effectively. Certain basic tools are involved: the development of a management information system for technical cooperation projects; diagnostic studies aimed at assessing current management practices and deficiencies; preparation of a policy paper (a Technical Cooperation Policy Framework Paper); and development of a technical cooperation programme that sets out priority technical cooperation projects—the equivalent on the technical assistance side of the now-widespread public investment programmes—for integration into the budget process.

Institutional requirements are the creation of a strong technical cooperation management unit—a focal point—as well as a political coordinating committee and added capacity at the operating agency level.

Comprehensive Programming

Many voices have begun to call for replacing the project-by-project approach to technical cooperation with programme approaches: technical cooperation and other inputs should be embedded in sectoral or subsectoral programmes, slices of which donors could finance. This would be a more rational approach and presumably make technical cooperation inflows easier to plan and manage. The growing support for the programme idea is indicated by U.N. resolutions in its favor, and wide donor consensus on this matter is reflected in the most recent DAC policy statement on technical cooperation:

> Increased emphasis should be given in the planning, selection and design of TC activities to a programme rather than a project-by-project approach. The programme approach should be based on thematic, sector-wide, multi-disciplinary and often multi-donor

actions.... [Aid agencies have agreed that they] ... will plan and manage ... aid increasingly in the context of coordinated support for larger sectoral programmes.... This principle should apply with particular force to technical cooperation. The effectiveness of technical cooperation has suffered from a piecemeal approach.[1]

4. Improve the Work Environment

If donors do better with the nuts and bolts of putting together and implementing technical cooperation projects, and if the expert-counterpart model is dropped and African governments succeed in improving their management of foreign assistance, technical cooperation still cannot be an effective vehicle for capacity building as long as other requirements are not met. Salary and other incentives remain derisory, and public sector work settings are uncongenial in other ways to hard work, job stability, and serious commitment to training. Without an appropriate work environment, technical cooperation for capacity building operates against heavy, perhaps insuperable obstacles. Yet this is the present reality: most African public sectors are in deep disarray, largely because of sharp declines in real wages but also because of general administrative deterioration and inadequate operating budgets.

Donors and African governments have responded in several ways. Their short-term response has been to expand greatly the practice of paying salary supplements to national staff as well as paying salaries of some local staff hired on aid-financed projects. Almost all donors oppose most of these payments in principle, because they oppose payment of any local costs and because of their disruptive side effects. But, in practice, donors pay supplements to keep their projects on track. The resulting side effects, unintended, have been highly disruptive: higher rates of job turnover and creation of productivity-reducing resentment on the part of civil servants who do not benefit from supplements or donor-financed salaries.

Medium- and long-term efforts have gone mainly into civil service reform and, in particular, reform of government pay and employment policies. More general administrative reform efforts have also been undertaken, but these are not numerous or highly intensive.

The pay and employment reforms have been and remain the centerpiece of the effort to improve work environments. The World Bank has been the lead player: between 1981 and 1989, the Bank financed 47 projects with significant civil service reform components in 20 African countries. UNDP has many such projects under way.

The salient feature of the Bank approach, which other donors have supported, is its emphasis on combined employment reduction and salary increases. The approach urges cutbacks in employment of lesser-skilled employees who are generally plethoric, and increases in the level of civil service wages, with the biggest boosts going to those at the upper end of the skill and wage ladder. This group includes managers and technicians, whose contributions are indispensable for any improvement in economic management but who are most disaffected because of low and declining real wages.

Payment of salary supplements by donors and donor employment of nationals on advantageous terms create deep dilemmas. On the one hand, government remuneration is normally so low that civil servants are forced to moonlight to make ends meet. Without donor hiring and payment of supplements, the willingness of national staff to participate in development activities as staff members or counterparts becomes problematic, with negative consequences for capacity building. Moreover, donors compete with one another in this area, so refusal of one or several donors to pay supplements and hire nationals cripples their programmes, while allowing other donors to move forward.

On the other hand, these payments are extremely disruptive. They lead to dual salary structures that cause resentments and probable withdrawal of effort by those who do not benefit. They distort incentives. Bidding for local staff often takes the form of disguised competition, which leads to increasingly convoluted forms of remuneration and additional distortions. Per diem payments, for example, become the main motive for staff travel, rather than staff interest or national need. Skills and energies are allocated more by the size and availability of donor payments than by national priorities.

Coherent donor proposals to address this corrosive and spreading problem are few. The 1991 DAC statement on new orientations for technical cooperation proposes that salary supplements should be avoided in principle, and used only in exceptional cases and in the following manner: they should be time-bound, follow explicit rules, and be fully transparent. They should be linked to civil service reform. And donor practices should be harmonized, possibly through a centrally administered mechanism created by joint agreement—perhaps a wages fund.

One recommendation for UNDP policy change would allow salary supplements only in the poorest countries and only if they are under clear fiscal duress, if no substitute measures are feasible and if, without supplements, project start-up or effectiveness would be seriously blocked. Only in-kind payments would be permitted (housing, transport, training, and the like) except in extraordinary circumstances. The payments would have to be for extra work, would be specified in project agreements, and would stop when the project ends.

A widely discussed idea (hinted at in the DAC paper cited above) is the creation of an extrabudgetary facility that would centralize all payments of salary supplements to government employees. Donor resources would be pooled in a common fund and donor coordination achieved through donor participation in the operation of this facility. This would be a stopgap measure, tied to civil service and general administrative reform.

PROBLEMS WITH THE CONSENSUS REFORM PROPOSALS

These proposals—to improve the design and implementation of technical cooperation projects; replace the resident expert-counterpart model with short-term advisers, local consultants, and institutional twinning; transfer management of technical cooperation to recipients; and reform civil service working conditions to make them more suitable for training and transfer of technology—are basically sound. If they were not, it is unlikely that they would be so widely recommended. But they have some weak points, particularly related to feasibility.

Proposals for Better Delivery of the Existing Technical Cooperation Package

Two questions arise in assessing these mostly sensible recommendations for better donor performance in designing and implementing technical cooperation projects. First, one of the key recommendations may not be correct. This is the insistence on tighter terms of reference and the use of measurable output indicators to judge performance of technical assistants. The "blueprint approach" that is implied runs counter to everyday experience. Consultants rarely find that terms of reference really fit the problems they discover when they arrive in-country. Terms of reference are quickly made irrelevant by swiftly changing events and, in any case, rarely provide much help in getting the job done. Blueprint-type terms of reference and the use of measurable performance measures are especially unsuitable for capacity-building technical cooperation projects because their principal outputs are soft and difficult to measure.

A more important problem is heavy reliance of these proposals on exhortation: donors "must" or "should" prepare projects more slowly, downgrade technical competence as a recruitment criterion, or do more and better supervision. But these deficiencies reflect deep-rooted habits of organizational behavior on the donor side, behaviors that are not easily changed. After all, many of the recommended reforms have been on the table for a decade or more and have been embraced by almost everyone. When reformers call for better performance in putting the nuts and bolts of technical cooperation projects together, they are really asking a lot of organizational leopards to change their spots. How and why these profound transformations can occur now is not clearly explained.

Changing the Mix of Delivery Modes

Replacing the expert-counterpart model with coaching arrangements, local consultants, and institutional twinning raises three problems. First, what is recommended is the replacement of a flawed but familiar mode of technical assistance delivery by new and potentially better models that are, however, untested in Africa. Pure coaching projects (one adviser who makes periodic visits) or modified coaching arrangements (when one resident expatriate recruits and manages short-term experts) are few; the expert-counterpart system remains dominant despite its critics. Use of local consultants for nonengineering kinds of services is only getting under way in most countries. And institutional twinning, despite its long history, is still sparse on the ground. For example, of 135 projects approved by the World Bank in 1987 and 1988, only 8 included definite plans for twinning.

Second, unanswered questions exist regarding the feasibility of these proposed approaches. With respect to coaching, unless local officials engaged in the activity are truly committed to it and able to resist diversions to other work, the pace of activity will slow when the coach goes home. Because the surrounding environment is always changing (cabinets are reshuffled, officials change jobs, droughts and civil disturbances erupt), work programmes require rapid recasting, which is not easily done from afar.

Also, use of intermittent coaches or other delivery systems relying on short-term visits by expatriate experts is extremely management intensive. It is therefore likely to require either strengthening of donor contracting and supervision capacities or broader use of consulting firms. It is not clear that intermittent coaches will be less costly on balance than use of resident advisers.

Third, the limited reliance on twinning, despite its many apparent advantages, suggests the existence of obstacles to its use. Lack of incentives to twin by industrial country organizations may be one such obstacle; profit-making organizations, after all, have to find advantages in twinning. Twinning also can be more expensive because aid agencies may have to find, train, and supervise institutions with little experience in developing countries.

Replacement of the expert-counterpart system, then, may not be as easy as it looks. Fresh thinking about donor administrative organization and many pilot schemes are needed to speed change and smooth the transition.

Transfer of Management Responsibility to Recipient Governments

Here two paradoxes prevail. First and most basic, those countries in which the technical cooperation process is most fully donor driven—countries where the need for national management is most intense—have the least capacity to do it. They therefore have to build the information base and the institutions that will allow effective management, notably a strengthened management agency (a focal point) to handle technical cooperation, analytic capacity for better policy making on technical cooperation, and ability to set priorities in the use of technical cooperation and to integrate it into the national budget process. But this creation of institutional capacity to manage technical cooperation must overcome the obstacles that make effective capacity building in general so difficult a task. At the very least, it is a long process that makes immediate and total transfer of responsibility for technical cooperation a high-risk operation.

The second paradox is that, despite agreements in principle voiced at international meetings, many donors are reluctant to hand over extensive technical cooperation management authority to recipient governments. Donors doubt that adequate capacity exists in many cases. They worry about corruption. They fear loss of control over their programmes and priorities, much of which is imposed by their headquarters and their national legislatures. They know that national management would require profound changes in the traditional ways of doing technical cooperation business.

All of this suggests that a gradualist reform strategy is needed. The highest-priority task, which is already under way in many countries as a result of NaTCAP exercises, is to generate better information about technical cooperation. Better information can have great potential impact, is a prerequisite for other reforms, and is relatively easy to introduce. Improvement in the quality of technical cooperation expenditures is another high-priority objective. It requires organizational and procedural changes that should receive early attention in any reform programme: the creation of a central coordinating entity, for example, and the strengthening of technical cooperation project vetting procedures.

Some commonly recommended steps should be given lower priority or done with less intensity than sometimes proposed. First, proposals to do comprehensive national assessments of technical cooperation needs should be resisted. They are often costly and time consuming and yield little more in the way of programme guidelines than more limited studies that focus on bottlenecks. In most cases, a light needs assessment is sufficient. Second, moves to technical cooperation programming should be gradual. It is unwise to try at the outset to produce a complete mirror image of public investment programmes. It is probably better to start with a shadow (unofficial) technical cooperation programme for a year or two and also to concentrate first on a few key sectors and ministries.

The adoption of programme approaches is often recommended as a means of simplifying the management of technical cooperation. It is an excellent idea in principle. But comprehensive programming requires more, not less, institutional competence than the existing project-by-project approach because comprehensive programming requires generalized ability to define sectoral strategies, priorities, and programmes in addition to preparation of better projects. Projects will remain the basic unit of action, with or without comprehensive programming.

Improvement of the Work Environment

Public sector salary structures and working conditions have to create renewed incentives if technical cooperation for capacity building is to become more effective. This requires at least two specific reforms: civil service rehabilitation that gives greater rewards for skill, effort, and responsibility; and containment of the destructive effects flowing from donor payment of salary supplements.

It would seem, on the surface, that these objectives would be easy to achieve. The reorganization of civil services, regrading of staff, and restructuring of salary arrangements to allow payment of incentive wages are well defined and straightforward tasks presenting few apparent obstacles to realization. And all that's needed to deal with the salary supplements problem is for donors to simply stop making these payments. After all, this would involve no more than adherence to their own announced policies, which in most cases rule out payment of supplements.

In practice, these targets have proved extremely elusive. The three-pronged World Bank approach (reduce civil service employment, raise average salaries, and increase salary differentials by granting bigger increases to higher-level employees) has had limited success. By 1989, removal of phantom staff, adherence to retirement rules, voluntary departure programmes, and similar schemes had led to reductions of employment in at least seven African countries. The biggest cuts were in Guinea and Ghana. By 1990, Ghana had cut its 1985 civil service employment by 16 percent, and Guinea by 23 percent. Some discharged employees reentered the public service in Ghana, however, diluting these gains.

The strategy of reform calls for generation of savings via staff cutbacks, which will permit later payment of higher wages to those who remain. But results are discouraging on this score. For 15 countries (most of them African) for which wage bill data are accessible and that had World Bank or IMF reform programmes, 10 had wage bill increases during the latter half of the 1980s.

Data on real salary increases and on skill differentials are equally discouraging. Average real civil service pay has risen in only a few countries since the mid-1980s. The pattern has been one of continuing decline in real salaries. Nor do available data show any tendency toward widening of skill differentials to favor higher-level employees. And subsidiary recommendations to increase direct wages and reduce allowances have rarely been followed.

The results of efforts to rationalize public sector salary structures are thus meager. And the prospects that this will change do not seem bright. One basic assumption on which efforts to reform salary structures are based is that African, and other low income, governments will cut employment of low-skilled, low-paid workers and hold down real wage increases of those who remain, while raising more appreciably the salaries of civil servants in the top grades. But few politicians, trade unionists, and intellectuals seem sympathetic to a policy that calls for greater relative sacrifice for the numerous and poor employees in the lower ranks of the civil service.

The call for a widening of skill differentials flies in the face of the egalitarian wage policies that have prevailed for more than 30 years throughout the continent. When making general salary adjustments to compensate for inflation, African governments have almost always favored lower-paid employees. Base rates for upper-scale posts have been stagnant for many years in most of the region. Nominal salary increases have rarely been across-the-board percentage rises. The policy of allocating wage increases in ways favorable to the less-well-paid workers clearly has deep historical, ideological, and political roots.

The outlook for broader reforms, the kind of general administrative overhaul needed to create a truly congenial environment for capacity building, is even less promising than that for civil service salary reform. The record of institution-building projects and administrative reform programmes worldwide is disheartening; academic and other assessments of past efforts at such reform vary from the polite "below expectations" to the more rude "dismal." Students of development management and administrative reform efforts differ about why this is so; some point to inappropriate reforms, others to faulty implementation strategies. But virtually all of them agree that administrative reform is extremely difficult, that risks of backsliding are high, that real success stories are few, and that in all cases it is a slow and long process.

The abolition or even the substantial reduction of salary supplements is not much easier. Donors and African governments face real dilemmas on this matter. Both are anxious that development programmes move forward. They recognize that low salaries and other deficiencies in public sector workplaces reduce productivity and encourage workers to seek additional forms of income. These considerations, plus donor pressures to spend, push them to create or tolerate ever-widening arrangements for salary supplements despite their devastating effects on civil service incentive structures and the demotivation they foment among nonfavored employees.

No satisfactory way out of these dilemmas has yet been discovered. The imposition of restrictions and conditions on supplements (such as those proposed by DAC and UNDP) is surely the right way to go. But comparable restrictions exist in principle now, and have not prevented the flowering of donor payments to nationals.

Specific proposals can be faulted on the grounds of desirability or feasibility. Thus, to require that governments commit themselves to continue special compensation provided by donors after project completion is poor policy and unrealistic: fiscal priorities may require dropping these payments and this kind of conditionality is not credible. The proposal to create a "foundation" or common resource pool from which salaries for strategic posts would be paid and salary supplements distributed (a Bolivian innovation) has as yet found no welcome in Sub-Saharan Africa. Few donors seem willing to surrender their freedom of action, and some raise doubts about the feasibility of the idea. It would require detailed control over their projects, and all donors would have to agree to the rules. Even if they adhered overtly, disguised competition could easily replace direct supplements with greater distortions resulting.

Nor have governments welcomed the common pool or common fund idea. Some officials argue that it would be a diversion from the main task of overhauling the civil service, and a source of diminished commitment to that objective. Also, the sparse results from a recent salary supplement scheme in Ghana are revealing: opposition from civil servants to overt payment of special supplements crippled this well-conceived effort.

Most concerned donors favor the proposal, set out in the 1991 DAC report on new orientations for technical cooperation already cited, to make payments of supplements conditional on general reform of the civil service. This is a safe and sound proposition. But it is not clear how meaningful it can be. Process conditionality of the kind implied is nearly impossible to monitor, and, anyway, civil service reform is necessarily a long stop-and-go process.

At present, there seems to be no convincing alternative to the shaky set of proposals for remedying the basic weaknesses in African work environments that have been addressed here. The amelioration of the current practice of paying salary supplements should be pursued. Clearer criteria should be set down for the hiring of national staff. Supplements should be time-bound and transparent, and greater coordination among donors should be encouraged. Experiments should be undertaken—for example, with the creation of special common pool arrangements for payment of local costs; schemes to attract exiles for key jobs in their home countries; and so-called enclave approaches, whereby a limited number of strategic agencies (such as customs departments and economic policy-making units) benefit from better salaries and other advantages.

With respect to criteria for hiring of nationals, good analytical and practical justification exists for donor financing of local salaries in some circumstances: minimization of project costs, relief from fiscal austerity that prevents use of locally available skills, and greater capacity-building impact. This means that formal and informal restrictions on the hiring of nationals for project design and implementation should be reduced or eliminated. Only one major restriction should prevail: that nationals can be financed only for employment on true projects, activities with a beginning and an end. To do otherwise is to undermine the objectives of a slimmer public sector and the sustainability of development-related activities.

Adoption of these kinds of measures will have positive effects. But it has to be admitted that there is no clearly lighted path to renewed public sector effectiveness. And this means that, at least for the medium term, public sector work environments will continue to be unfriendly to sustainable capacity building and therefore inhospitable to technical cooperation.

ADDITIONAL (NONCONSENSUS) RECOMMENDATIONS

Numerous question marks thus continue to surround the consensus proposals for enhancing the capacity-building impacts of technical cooperation. This does not diminish their basic soundness and desirability. But it does give urgency to the search for additional or back-up measures that address neglected aspects of the problem and bolster the prospects for successful reform.

We focus on three such measures. These are not much discussed in the literature and are not part of the consensus package. Each has its own question marks and problems of implementation. But each seems implementable and should raise the capacity-building effectiveness of future technical cooperation.

Gap-Filling Without Shame

Specialized skills continue to be scarce in most African countries, and it will be some time before these are filled. About a third of the 300 medical doctors in Burundi are foreign, almost all in operational jobs. In the natural and social sciences, agricultural research, specialized engineering occupations, and statistics, economics, and similar fields, national capacity remains thin in much of the continent.

These posts would be best filled by recruiting operating staff—that is, using substitution or gap-filling technical assistants who would occupy line posts and hence be integrated into the national administration. The pretense about counterparts would be dropped, explicitly and frankly. What would be involved is a return to the system of topping-off salaries of expatriates in operational jobs once widely used by the British and the United Nations. This would simply ratify existing reality because many expatriates imported through technical cooperation programmes to work as advisers in fact work as line staff.

Bringing these technical assistance personnel out in the open in this way would have several advantages. It would integrate them better into the existing hierarchical job structures. It would be clearly established that they are members of a team. They would work with nationals as peers and have a normal status, subordinate to some coworkers and supervisor of others. That ambiguous status of adviser with counterpart would disappear.

Also, the determination of personnel needs would be somewhat simplified. The existence of unfilled established posts signals a possible need for help. And the fact that the post is established makes sustainability more likely. Fiscal crises with accompanying hiring freezes and the need to cut back on employment in the public sector do weaken this argument somewhat, because governments will continue to turn to expatriate personnel to fill posts that nationals could fill. But the number of posts to be filled this way would be small and declining. And adoption of gap-filling would facilitate transfer of technical cooperation control to recipient governments.

Empirical evidence suggests in any case that this system would be more effective and easier to manage. Countries like Botswana and Papua New Guinea, which are often cited for their good management of technical cooperation, have long relied on just these arrangements.

Introduction of Market Elements

The technical cooperation "market" or system has several features that induce dysfunctional bureaucratic behavior and contribute significantly to the poor results now so widely deplored. First and most important, costs and prices play only a small role in determining the supply of and demand for technical assistance personnel. Opportunity cost consciousness is stunted, particularly on the demand side. This encourages the economically irrational use of aid resources.

Second, "buyers" or takers of technical cooperation have limited flexibility in choosing the mix of inputs they want; technical cooperation packages are bundled by donors (suppliers) and cannot be unbundled by recipients (demanders). This leads recipients to request or to accept passively more technical assistance personnel than they want, simply to get the fringe benefits—vehicles, computers, copiers, supplies, and access to training and travel abroad. Moreover, donor practices restrict the right of recipients to choose between local and imported skills, even when local skills are available, would be cheaper to use, and can be suitably employed in donor-financed projects.

The market for technical assistance personnel is thus extremely imperfect. Indeed, prices are largely unseen. Governments believe they incur few or no costs when they ask for or accept technical assistance that is grant financed, as most is. Opportunity costs—mainly in the form of reduced inflows of other, substitutable forms of foreign assistance—are uncertain, and recipients perceive them to be small. And whatever are the financial and opportunity costs to the recipient country, the host agency, the operating entity that is the decision-making unit on the demand side, has no incentive to economize on its use. In fact, the incentives are perverse: users pay nothing but enjoy the benefits, direct and indirect, from the presence of technical assistance personnel in their agency.

Other distortions exist on the supply side. Donors have many reasons to urge technical assistance personnel on host governments and few to abstain. In addition, the tied package of expatriate personnel, equipment and supplies, and training induces irrational use of these resources. It is no surprise, then, that technical cooperation projects proliferate, local commitment to them is minimal, and the use of technical assistance personnel is often out of line with national priorities.

Two basic policy reforms are called for: the introduction of greater cost consciousness and untying of the package. The former can be encouraged by institutional changes and by exhortation. But effective introduction of cost consciousness will require that a price be attached to all technical assistance personnel, this price to be paid out of the user agency's budget and not from a global allocation in the budget of the finance ministry. And governments should be allowed to separately request personnel, equipment and supplies, and training.

Illustrative of the way cost awareness can be developed by institutional change is the introduction of technical cooperation programmes. The simple setting out of the full array of technical cooperation projects, with their costs and sources of financing, is an eye-opener for all concerned. In several countries, technical cooperation commissions are doing ex ante evaluation of projects, forcing agencies to justify their requests. In at least one country (Guinea), effective programming has led to cuts in spending on technical cooperation.

These positive effects, caused by greater knowledge about the magnitude and costs of technical assistance personnel, would be magnified many times over by the introduction of prices at two levels. At the macroeconomic level, a ceiling has to be placed on technical cooperation budgetary allocations; technical cooperation should be made to compete for available foreign aid and domestic resources along with other forms of aid and other claimants on domestic revenues. On the microeconomic level, a price should be exacted from user agencies for each person-month of technical assistance personnel they use.

A first step, then, which has already begun in some countries, is to extend the technical cooperation programme concept by making the costed inventory of existing technical cooperation projects the basis of a national technical assistance budget. Within this budget, user agencies would have an allocation based on their programme submissions. The budget would be the equivalent of the popular public investment programmes. The total size of the annual allocation for technical assistance personnel, and its sectoral composition, would be determined by weighing technical cooperation needs against competing demands for the public investment programme, for ongoing operations and maintenance and for the operation of completed projects coming on-stream.

The idea of charging a fee to user agencies for their technical assistance personnel is not new; it is the equivalent of local counterpart contributions that many donors require in capital projects. But it is far from current practice and runs counter to some trends: the World Bank, for example, which normally makes clients pay for their technical assistance, created in 1992 a grant facility for certain kinds of technical cooperation projects.

With respect to the unbundling of the technical cooperation package, donors have to allow freer choice in the use of technical cooperation resources. In particular, recipients should not be induced to take expatriate experts because they want the training, equipment, and supplies that come with them. At the same time, donors and recipients should begin the search for longer-term solutions and launch a direct and more comprehensive attack on the problem of inadequate public sector support services. It is not acceptable that all the actors continue to tolerate the extraordinary spectacle of so many governments trying to operate in the absence of the most elemental tools of modern governance.

Much more money and imagination should therefore go into freestanding capacity-building technical cooperation programmes aimed at devising practical solutions for problems such as lack of means of transport, nonexistent maintenance systems for computers and copiers, or withering of in-service training systems. Privatization of these kinds of services may present new possibilities for effective action; these are activities that lend themselves readily to contracting-out to private agents. In any case, more fundamental and comprehensive reform efforts should replace the existing piecemeal, institutionally corrosive arrangements whereby resources for support services come attached to resident expatriate experts.

Training should also be made more autonomous. Only the on-the-job components should remain attached to the technical assistance personnel. The scattered training inputs attached to technical cooperation projects, such as scholarships and seminars, should be integrated into more comprehensive training strategies.

Many details remain to be worked out to ensure a viable "marketization" of technical cooperation, and to overcome or avoid the numerous obstacles and pitfalls to make it work properly. But potential gains are big. The structure of incentives in the market would be made to work for rational behavior rather than against it. If successful, this approach would vastly facilitate overall reform.

Even a mildly successful effort—perhaps even a failed one—could have positive impacts. It would reinforce the movement toward more fully programmed technical cooperation. It would increase awareness among user agencies of the need to cost and programme their planned technical cooperation needs properly. The shifting of payment burdens directly to user agencies would encourage greater transparency of costing and the adoption of simplified, more efficient systems of payment to technical assistance personnel—shifting from in-kind housing subsidies to cash allowances, for example.

Because of the uncertainties and risks that such reforms entail, the transition should be gradual and tentative. It might be desirable to start with a few key sectors or ministries, extending the process over time. And the process might begin with free-standing technical cooperation projects because the pricing and unbundling principles would be easier to apply in this type of project.

Administrative Reform and Downsizing the State

Most of the consensus propositions on administrative reform are sensible. The best path to a satisfactory enabling environment for technical cooperation is for donors and governments to focus on building stronger institutions in the public sector. Faster rehabilitation of public sector management, particularly civil service reform, is indispensable. Public sector salaries have to be made more conducive to job commitment and acceptance of responsibility. Personnel management has to be revitalized.

Two proposals can be added to this agenda for reform. The first consists of recommendations for amelioration of present reform strategies or approaches; the second is to build contingencies into the reform process itself in case of failure. Neither of these ideas is original; indeed they are already present in some reform programmes.

Administrative reform efforts could be improved by:
- Evaluating more intensively the efficacy of specific instruments, such as rolling public investment programmes and public enterprise performance contracts, and using lessons of experience more effectively;
- Giving civil service reform an even higher priority than it now has in the allocation of foreign assistance, especially technical assistance;
- Downplaying somewhat the job retrenchment focus of current reform strategies because its results are rarely substantial, it makes contentious the whole civil service reform programme, and the problem of overstaffing may be reduced over time by attrition and by indirect attack;

- Giving more attention to aspects of administration other than personnel—improvement of supplies management and support services, for example;
- Distancing administrative reform from externally imposed conditionalities, which dilute local ownership; and
- Giving greater attention to partial approaches, to piecemeal reform—notably, by creating reform enclaves or islands that would improve performance of a few critical functions or agencies, such as customs administration, tax collection, and economic policy analysis.

Even with these improvements in the way administrative reform is tackled, the enterprise will remain high risk and long term for reasons outlined earlier. Back-up strategies are therefore prudent, even essential. In the short run, for example, it makes little sense to rely on expert-counterpart models of technical cooperation delivery when the prospects for salary reform are low, which means that slack job commitment and high turnover are likely to continue. And even were salary reform to come, the remaining institutional weakness in public sectors would continue to hobble capacity-building efforts.

Technical cooperation providers and African governments thus have to do their planning and design their programmes on the assumption that over the next 10 to 15 years public sectors are almost sure to remain in fiscal duress, work incentives will remain poor, and the operational efficiency of government agencies will remain low. Donors and governments should therefore follow a parallel track even while undertaking major efforts at administrative reform. They should find alternative ways to increase and improve the provision of vital government services.

These alternatives are at hand. These involve mainly lightening the public sector burden by encouraging other agents to do more—private profit seeking entities, semipublic bodies, or nonprofit voluntary agencies. The instruments are well known: deregulation (getting government out of the way of private service providers), contracting-out, and internal divestiture—the spinning off of activities that are not central to the mission of government.

These approaches have great potential, still largely unexploited, to reduce public sector burdens and increase the supply of public services and the efficiency of their delivery. Services such as data processing, all kinds of maintenance, research and policy analysis, agricultural input delivery, branching and billing, and training and maintenance for public utilities are only a few of the functions or activities that are amenable to contracting-out.

Internal load-shedding can allow government agencies to concentrate on their core responsibilities. Private and semiprivate foundations or institutes can be formed to perform research and analysis; they can create salary incentives and other inducements for productive and sustainable performance. More complete deregulation in strategic sectors, in education, health, trade, and transport, for example, would surely increase supply and efficiency of delivery. Encouragement of local consulting capacity along lines suggested earlier could be accelerated by nurturing measures such as smaller contract awards and encouragement of twinning with foreign consulting firms. National management burdens could be reduced by creating stronger, better-used regional centers of excellence.

More technical cooperation resources should be allocated to the achievement of these approaches to state shrinkage. Many benefits can be anticipated, among them reduced public sector wage bills, greater entrepreneurship and supply responsiveness, and increased productivity via specialization.

Two benefits from the downsizing of the state are particularly pertinent from the perspective of building durable capacity in economic management. Privately based organizational competence is not likely to disappear as foreign assistance diminishes in the future. And efforts to reform public sector economic management would be made simpler and more manageable, and hence would be more likely to yield durable results.

All the recommendations set out here have imperfections, as do most proposed solutions for real world problems. But the analysis in this volume has made clear that radical changes are necessary if technical cooperation is to be put on the right track. Africa will become more self-reliant one way or another. All of its people will benefit if self-reliance comes in an environment of renewed economic growth, because only in such an environment can their energies and potentials be fully nurtured and mobilized.

More capital investment is certainly needed for faster growth. But better-trained people and stronger institutions are needed even more. And that is what technical cooperation for capacity building is all about. The problem, which has been underscored repeatedly in this book, is that the inherited technical cooperation processes and instruments for building capacity in economic management are profoundly flawed. Unless overhauled, technical cooperation will remain ineffective in transmitting new technology and creating new human capital—ineffective, that is, in building Africa's capacity to manage its own affairs.

The decline in the availability of aid gives reform of technical cooperation special urgency. Real net inflows of official development assistance to Sub-Saharan Africa have declined for two successive years (1990 and 1991). Important donors are cutting aid budgets or reallocating to the post-Socialist states. It is more than ever essential to use African foreign assistance resources efficiently. And that large part of total assistance to Africa that is made up of technical cooperation expenditures can unquestionably be used more effectively and efficiently. More capacity building can surely be done with less money.

The stakes are therefore large. One main vehicle for economic change and growth, a central pillar in the whole structure of development assistance and a major user of development assistance money, is in need of basic repairs. It is up to African governments and their foreign partners to respond quickly and vigorously to this need. Not many dimensions of development policy are more critical or more deserving of sustained attention because not many concern so closely the creation of basic skills and institutions required for faster economic growth and greater self-reliance. In few areas of policy are the costs of inaction or misguided action more far reaching.

NOTE

1. OECD, DAC, "Principles for New Orientations for Technical Cooperation," 1991, pars. 27-28.

ANNEX

STATISTICS ON TECHNICAL COOPERATION

EXPLANATORY NOTES FOR STATISTICAL TABLES

The following tables show overall trends in technical cooperation to Sub-Saharan Africa. They are based on OECD, World Bank, and UNDP sources. OECD data on aid flows constitute the only source of technical cooperation data that is consistent, has wide country coverage, and covers a significant time period. At the country level, only a few countries have reliable data on technical cooperation resources.

These data should be interpreted with caution because of the following drawbacks:

- OECD data on technical cooperation cover only free-standing grants provided by DAC member countries and the multilateral agencies. The data thus do **not** include (1) loan-financed technical cooperation activities, (2) investment-related technical cooperation, (3) technical cooperation provided by nongovernmental organizations, or (4) technical cooperation provided by Eastern European countries and other developing countries; and

- The national statistics that provide the inputs for OECD data sometimes use different definitions and different fiscal years. The consolidated data thus have some uncertainties.

Classification of Countries

Sub-Saharan Africa includes all countries located south of the Sahara, with the exception of the Republic of South Africa. Namibia is not included because of absence of data.

The category of the Least Developed Countries (LDCs) includes the 28 African countries that were classified as such by the United Nations in 1989, specifically: Benin, Botswana, Burkina Faso, Burundi, Cape Verde, Central African Republic, Chad, Comoros, Djibouti, Equatorial Guinea, Ethiopia, Gambia, Guinea, Guinea-Bissau, Lesotho, Malawi, Mail, Mauritania, Mozambique, Niger, Rwanda, São Tomé and Principe, Sierra Leone, Somalia, Sudan, Tanzania, Togo, and Uganda. For consistency, Liberia has been excluded from this list because it was granted LDC status only at the end of 1989. Angola and Senegal, which are considered as "as if" LDCs but which do not formally belong to the group, have also been excluded. In each table, the 45 African countries considered are listed in alphabetic order.

> The symbol "-" in the tables indicates nil, negligible, not available, or not applicable.

Table 1. Net Disbursements of Official Development Assistance by Recipient Country, 1970-1989

Data are from OECD, *Geographical Distribution of Financial Flows to Developing Countries*, Paris, various years.

Official development assistance (ODA) is defined by OECD as those flows to developing countries and multilateral institutions provided by official agencies, including state and local governments, or by their executive agencies, each transaction of which meets the following test:

(1) It is administered with the promotion of economic development and welfare of developing countries as its main objective; and

(2) It is concessional in character and contains a grant element of at least 25 percent.

The data include net disbursements of official development assistance to Sub-Saharan African countries from the following countries and institutions:

- Bilateral assistance from member countries of the OECD/DAC;
- Multilateral assistance from:
  — The European Community;
  — U. N. specialized agencies, World Bank, African Development Bank and Fund, and International Fund for Agricultural Development; and
  — Arab member countries of OPEC and Arab-financed multilateral agencies.

Resources provided by other developing countries, by Eastern European countries, and by nongovernmental organizations are not included in the table.

Table 2. Technical Cooperation Grants by Recipient Country, 1970-1989

Source: Same as for Table 1. Data refer to net disbursements.

Technical cooperation is defined by OECD as activities whose primary purpose is to augment the level of knowledge, skills, technical know-how, or productive aptitudes of the populations of developing countries—that is, to increase their stock of human intellectual capital or their capacity for more effective use of their existing factor endowment. These data cover only a part of total technical cooperation flows: free-standing technical cooperation provided on grant terms by DAC countries. They therefore do **not** include three kinds of technical cooperation:

- Technical cooperation provided by non-DAC member countries, including Arab countries and nongovernmental organizations;
- Technical cooperation provided as part of investment projects; and
- Technical cooperation financed by loans.

Definitions and concepts of technical cooperation are discussed in detail in Chapter Two.

Table 3. Country Allocation of Technical Cooperation Grants from Major Donors, 1970 and 1989

Source: Same as for Table 1.

Table 4. Country Allocation of Technical Cooperation Grants from Major Donors in Percentages, 1970 and 1989

Table 5. Technical Cooperation Grants to Sub-Saharan Africa: Distribution by Major Donor Sources in Percentages, 1970 and 1989

Based on Table 3, this table shows for each recipient country the major sources of technical cooperation as a percentage of total technical cooperation resource inflows.

Table 6. Distribution of Technical Cooperation Grants by Principal Economic Sectors in 1988 (%)

Data are extracted from UNDP Development Cooperation Reports, prepared annually by the UNDP Field Offices.

Table 7. Economic Significance of Technical Cooperation Grants for Recipient Countries

Sources of data are the following:
    7.1   GNP per capita: World Bank, *World Bank Atlas*, Washington, D.C., various years.

7.2 Technical cooperation expenditures per capita: OECD.

7.3 Technical cooperation expenditures as percentage of GDP: OECD for technical cooperation and World Bank for GDP.

7.4 Technical cooperation expenditures as percentage of ODA: OECD.

7.5 Technical cooperation expenditures as percentage of government revenue: OECD for technical cooperation and UNDP/World Bank (*African Economic and Financial Data*, 1990) for government revenue.

7.6 Technical cooperation expenditures as percentage of export earnings: OECD for technical cooperation and World Bank for exports (World Tables, 1989-1990).

7.7 Technical cooperation grants per capita, rank ordering, 1989.

Government revenue refers to current revenues (tax and non-tax) and capital revenue, such as proceeds from the sale of real assets, including lands. They do not include grant receipts from other governments or international organizations.

Averages calculated for Sub-Saharan Africa as a whole and the LDC group are population-weighted averages.

Annex: Statistics on Technical Cooperation 279

LIST OF ANNEX TABLES

Table		Page
1	Net Disbursements of Official Development Assistance by Recipient Country, 1970-1989	281
2	Technical Cooperation Grants by Recipient Country, 1970-1989	283
3	Country Allocation of Technical Cooperation Grants from Major Donors, 1970 and 1989	285
4	Country Allocation of Technical Cooperation Grants from Major Donors in Percentages, 1970 and 1989	288
5	Technical Cooperation Grants to Sub-Saharan Africa: Distribution by Major Donor Sources in Percentages, 1970 and 1989	291
6	Distribution of Cooperation Grants by Principal Economic Sectors in 1988 (percentage)	294
7	Economic Significance of Technical Cooperation Grants for Recipient Countries	296
	7.1 GNP per Capita	296
	7.2 Technical Cooperation per Capita	296
	7.3 Technical Cooperation as Percentage of GDP	297
	7.4 Technical Cooperation as Percentage of ODA	297

7.5 Technical Cooperation as Percentage of Government Revenue   298

7.6 Technical Cooperation as Percentage of Exports   298

7.7 Technical Cooperation Grants per Capita, Rank Ordering, 1989   299

ANNEX: Statistics on Technical Cooperation 281

TABLE 1: NET DISBURSEMENTS OF OFFICIAL DEVELOPMENT ASSISTANCE BY RECIPIENT
COUNTRY, 1970-1989 (in millions of U.S. dollars)

COUNTRY	1970	1971	1972	1973	1974	1975	1976	1977	1978	1979
1 Angola	0.0	0.0	0.1	0.2	0.4	4.8	38	48	47	47
2 Bénin	15	29	20	27	33	54	55	49	62	85
3 Botswana	14	18	32	37	37	51	48	48	69	100
4 Burkina Faso	22	29	34	57	97	89	84	110	159	198
5 Burundi	18	22	26	27	38	48	45	48	75	95
6 Cameroun	59	48	63	61	62	125	134	176	178	277
7 Cap-Vert						8.8	25	27	36	33
8 Comores	7.9	8.8	10	18	27	22	26	12	13	18
9 Congo	16	17	23	27	38	56	73	49	81	91
10 Côte d'Ivoire	53	51	48	64	76	101	108	106	131	162
11 Djibouti	12	11	13	20	29	34	28	46	100	23
12 Ethiopia	40	47	48	67	121	135	141	116	140	191
13 Gabon	24	24	27	39	25	63	34	28	44	37
14 Gambia	1.3	3.6	5.0	6.5	10	8.1	12	22	36	37
15 Ghana	59	57	59	41	37	126	64	91	114	169
16 Guinée	10	10	5	11	30	15	12	22	60	56
17 Guinée-Bissau	0.1	0.2			3.3	19	23	38	50	53
18 Guinée-Eq.				1.0	16	2.2	0.4	0.9	0.6	2.7
19 Kenya	58	67	72	96	118	129	162	163	248	351
20 Lesotho	10	17	14	14	21	30	30	39	50	66
21 Liberia	13	13	13	11	16	21	27	34	48	81
22 Madagascar	48	47	55	53	63	85	63	61	91	138
23 Malawi	37	32	36	30	42	64	63	79	99	142
24 Mali	21	30	38	71	118	145	89	113	163	193
25 Mauritanie	7.5	12	9.1	31	88	59	168	160	238	167
26 Mauritius	6.1	9	11	14	25	29	17	22	44	32
27 Mozambique	0.1	0.1	0.1	0.1	0.7	22	72	80	105	146
28 Niger	32	38	43	71	137	141	129	97	157	174
29 Nigeria	108	107	83	77	73	82	53	43	43	27
30 République Centrafricaine	14	16	26	26	37	57	38	42	51	84
31 Rwanda	22	25	30	39	47	91	79	96	125	148
32 Sao Tomé and Principe						0.9	12	3.1	4.1	3.1
33 Sénégal	43	53	48	79	139	133	127	123	223	307
34 Seychelles	4.0	7.7	8.7	8.0	8.5	7.5	7.4	11	17	25
35 Sierra Leone	6.9	11	10	14	11	18	15	26	40	54
36 Somalia	28	31	24	36	40	73	105	256	207	181
37 Sudan	6.4	11	37	42	97	127	369	231	318	571
38 Swaziland	6.1	2.2	9.0	11	16	16	15	29	45	51
39 Tanzania, U.Rep. of	51	62	61	100	163	295	268	340	424	589
40 Tchad	22	31	31	45	79	65	62	83	125	86
41 Togo	17	19	22	26	39	42	43	64	103	110
42 Uganda	33	32	30	15	30	39	25	22	23	47
43 Zaïre	89	109	123	140	181	205	194	261	317	416
44 Zambia	13	22	22	46	58	87	62	109	185	278
45 Zimbabwe	0.7	0.8	0.9	0.9	1.6	4.0	6.3	6.7	9.2	13
LDCs	449	543	604	830	1389	1754	2064	2271	3032	3650
SSA	1049	1179	1269	1596	2324	3028	3249	3630	4894	6151

TABLE 1: 1980-1989—Continued

COUNTRY	1980	1981	1982	1983	1984	1985	1986	1987	1988	1989
1 Angola	53	61	60	75	95	92	131	135	159	147
2 Bénin	91	82	81	86	78	96	138	138	162	248
3 Botswana	106	97	102	104	103	97	102	156	151	160
4 Burkina Faso	212	217	213	184	189	197	294	281	298	282
5 Burundi	117	122	127	140	141	143	188	202	188	198
6 Cameroun	265	199	212	129	187	159	225	213	285	469
7 Cap-Vert	64	50	55	60	64	70	109	88	87	75
8 Comores	43	47	39	38	41	48	46	54	52	45
9 Congo	92	81	93	108	98	71	110	152	89	89
10 Côte d'Ivoire	210	124	137	156	128	125	187	254	439	412
11 Djibouti	72	64	59	66	102	82	115	106	93	76
12 Ethiopia	212	245	200	339	364	715	636	634	970	742
13 Gabon	56	44	62	64	76	61	79	83	101	133
14 Gambia	55	68	48	42	54	50	101	100	82	93
15 Ghana	193	145	141	110	216	203	372	373	474	553
16 Guinée	90	107	90	68	123	119	175	213	262	348
17 Guinée-Bissau	60	65	65	64	55	58	71	107	99	102
18 Guinée-Eq.	9.3	10	14	11	15	17	22	43	43	42
19 Kenya	397	449	485	401	411	438	456	572	807	972
20 Lesotho	94	104	93	108	101	94	88	108	108	128
21 Liberia	98	109	109	118	133	91	97	78	65	59
22 Madagascar	230	234	242	184	153	188	316	322	304	323
23 Malawi	143	138	121	117	159	113	198	281	366	399
24 Mali	267	230	210	215	321	380	372	366	427	455
25 Mauritanie	176	214	187	176	175	209	225	182	184	244
26 Mauritius	33	58	48	41	36	29	56	65	59	59
27 Mozambique	169	144	208	211	259	300	421	651	893	768
28 Niger	170	193	258	175	161	305	307	353	371	297
29 Nigeria	36	41	37	48	33	32	59	69	120	345
30 République Centrafricaine	111	102	90	93	114	104	139	176	196	193
31 Rwanda	155	154	151	149	165	181	211	245	252	235
32 Sao Tomé and Principe	3.9	6.1	9.9	12	11	13	12	17	24	33
33 Sénégal	262	398	285	322	368	294	567	641	568	651
34 Seychelles	22	17	18	16	15	22	29	24	21	20
35 Sierra Leone	91	63	8	66	61	66	87	68	102	101
36 Somalia	447	374	462	343	350	353	511	580	433	421
37 Sudan	620	681	740	962	622	1129	945	898	947	752
38 Swaziland	50	37	28	34	30	25	35	45	38	30
39 Tanzania, U.Rep. of	679	703	684	594	558	487	681	882	982	920
40 Tchad	35	60	65	95	115	182	165	198	264	242
41 Togo	91	63	77	112	110	114	174	125	199	183
42 Uganda	114	136	133	137	163	182	198	279	363	405
43 Zaïre	428	394	348	315	312	325	448	627	576	638
44 Zambia	318	232	317	217	240	329	465	430	478	390
45 Zimbabwe	164	212	216	209	298	237	225	294	273	266
LDCs	4498	4533	4662	4765	4771	5901	6731	7530	7918	7538
SSA	7403	7367	7501	7310	7599	8621	10586	11906	13221	13744

Source: *Geographical Distribution of Financial Flows to Developing Countries*, OECD, Paris, various years.

TABLE 2: TECHNICAL COOPERATION GRANTS BY RECIPIENT COUNTRY, 1970-1989
(in millions of U.S. dollars)

COUNTRY	1970	1971	1972	1973	1974	1975	1976	1977	1978	1979
1 Angola	0.1	0.1	0.0	0.1	0.1	0.2	5.7	18.7	10.6	14.4
2 Bénin	6.0	10.0	8.3	10.2	12.4	20.0	16.0	13.7	16.9	20.4
3 Botswana	2.5	4.8	4.5	7.8	9.5	13.7	15.5	18.0	23.0	36.3
4 Burkina Faso	8.2	9.5	12.8	15.3	19.5	32.5	37.5	37.6	49.1	60.7
5 Burundi	10.5	11.4	13.6	15.5	17.1	23.1	23.0	25.6	30.4	35.7
6 Cameroun	17.2	18.4	24.7	31.1	33.1	48.7	47.8	47.9	51.9	64.0
7 Cap-Vert										
8 Comores	4.8	6.4	3.0	8.6	13.4	7.0	2.2	2.3	1.3	2.9
9 Congo	10.1	11.5	11.8	12.8	14.8	23.4	25.4	25.5	27.9	33.9
10 Côte d'Ivoire	20.7	23.8	26.8	35.7	39.4	55.2	54.1	57.3	61.6	83.7
11 Djibouti	6.2	8.1	5.1	9.6	13.7	11.1	14.0	15.2	14.3	19.0
12 Ethiopia	19.6	20.9	22.2	28.5	34.6	33.3	31.5	30.3	26.5	29.2
13 Gabon	6.8	8.0	10.2	13.0	13.0	19.6	23.9	24.3	28.1	24.4
14 Gambia	0.6	1.1	1.6	1.9	2.1	2.6	4.3	3.3	6.8	9.6
15 Ghana	14.9	15.9	14.9	17.6	19.7	25.9	27.5	29.2	37.5	35.4
16 Guinée	4.2	4.2	1.0	1.1	1.3	3.2	5.5	6.0	10.6	11.1
17 Guinée-Bissau									6.7	9.0
18 Guinée-Eq.										
19 Kenya	27.1	32.5	41.4	45.6	49.9	58.7	69.5	66.8	81.3	103.1
20 Lesotho	2.5	2.8	3.2	5.0	6.5	8.7	12.0	12.7	16.3	19.9
21 Liberia	8.0	11.4	9.8	10.4	10.5	12.6	10.9	12.3	15.9	19.2
22 Madagascar	20.8	21.0	26.3	21.9	23.5	33.0	29.2	25.7	24.8	37.5
23 Malawi	8.9	9.8	11.1	11.6	14.6	16.3	17.4	17.6	24.0	30.0
24 Mali	8.1	9.2	11.9	12.9	18.9	24.4	23.2	25.6	35.4	49.5
25 Mauritanie	4.4	4.6	6.4	7.8	10.6	11.4	13.3	13.8	18.6	24.5
26 Mauritius	1.4	2.2	3.1	4.6	5.1	6.5	6.2	5.8	7.8	9.6
27 Mozambique	0.0	0.0	0.0	0.0	0.3	4.1	7.3	12.3	26.9	33.2
28 Niger	11.2	12.0	14.5	14.0	24.4	29.1	27.9	30.5	37.4	47.3
29 Nigeria	36.4	31.4	37.9	44.2	40.3	40.8	31.2	31.6	37.2	41.5
30 République Centrafricaine	8.2	8.3	12.2	12.1	12.9	19.2	19.4	20.8	24.6	32.9
31 Rwanda	11.8	13.7	16.2	19.3	23.4	31.5	32.3	41.2	42.8	50.9
32 Sao Tomé and Principe										
33 Sénégal	20.7	20.0	27.6	35.6	41.6	63.1	57.4	55.4	82.8	99.3
34 Seychelles	0.6	0.9	1.0	1.7	1.9	2.4	2.5	2.9	4.8	5.8
35 Sierra Leone	5.3	4.9	4.2	5.6	4.2	7.8	10.4	10.4	13.2	14.9
36 Somalia	10.5	10.2	9.7	12.0	12.7	19.4	16.0	19.4	21.5	32.1
37 Sudan	6.4	6.1	21.6	19.2	15.5	28.0	30.9	37.7	60.3	69.1
38 Swaziland	1.9	2.2	3.8	5.3	7.1	9.1	10.5	9.7	10.5	13.1
39 Tanzania, U.Rep. of	21.3	23.8	34.7	41.3	47.6	60.2	76.7	90.2	106.0	138.2
40 Tchad	10.0	12.3	15.0	20.1	19.6	25.3	22.0	24.6	29.2	21.4
41 Togo	7.4	9.1	11.1	13.3	14.8	17.0	16.9	18.9	21.8	25.0
42 Uganda	14.4	17.8	16.1	11.9	7.9	9.1	8.4	7.7	12.0	16.4
43 Zaïre	41.9	52.8	65.4	78.6	93.5	110.0	97.6	107.9	145.0	160.1
44 Zambia	13.2	19.1	21.7	25.5	28.2	34.3	39.1	41.8	52.1	71.5
45 Zimbabwe	0.7	0.8	0.9	0.7	1.4	4.0	6.1	6.7	8.2	12.3
LDCs	193	221	260	305	358	458	484	535	676	839
SSA	436	493	587	689	781	1006	1003	1079	1364	1668

TABLE 2: 1980-1989—Continued

COUNTRY	1980	1981	1982	1983	1984	1985	1986	1987	1988	1989
1 Angola	15.7	18.1	19.7	25.5	23.6	25.3	28.4	34.5	43.6	47.1
2 Bénin	25.9	25.1	27.9	25.5	28.1	30.3	38.4	48.3	44.9	47
3 Botswana	47.5	48.0	42.8	40.5	34.8	36.9	43.4	49.8	63.8	57.7
4 Burkina Faso	73	72.0	73.7	65.2	71.4	69.7	96.2	99.9	99.0	94.8
5 Burundi	45	44.2	46.9	43.3	42.7	45.7	53.8	54.3	53.9	51.6
6 Cameroun	84.7	76.6	62.8	59.2	64.7	62.5	91.7	92.6	118.0	106.7
7 Cap-Vert					17.0	19.1	24.5	28.4	24.7	17.1
8 Comores	7	10.1	9.6	9.9	9.4	11.1	13.6	20.4	16.3	17.1
9 Congo	37.1	30.7	30.7	32.2	34.4	31.1	38.5	48.6	38.3	36.8
10 Côte d'Ivoire	99.8	62.8	56.1	84.8	59.1	43.5	87.1	79.3	64.9	99.4
11 Djibouti	27.8	29.8	30.9	29.3	29.8	29.7	45.5	37.9	38.7	33.9
12 Ethiopia	44.2	63.9	53.1	63.6	80.9	104.2	115.5	146.9	195.1	218.1
13 Gabon	37.6	28.0	28.4	24.1	29.9	30.2	33.1	40.3	37.7	31.6
14 Gambia	12.7	13.8	17.4	15.8	14.8	17.9	32.5	21.5	26.7	23.5
15 Ghana	42.4	42.3	34.6	29.1	25.3	30.6	39.0	46.1	58.9	59.4
16 Guinée	18.7	21.4	20.5	14.4	17.9	19.9	26.5	35.8	39.9	43.2
17 Guinée-Bissau	12.1	12.7	15.2	16.1	11.8	14.9	22.5	27.3	29.5	25.2
18 Guinée-Eq.					6.7	6.0	7.8	6.6	10.1	12
19 Kenya	127.9	127.8	116.8	122.1	109.5	116.9	156.3	153.8	176.9	174.7
20 Lesotho	29.1	31.2	34.7	34.7	34.0	31.3	39.1	38.8	47.2	46.6
21 Liberia	23.8	24.1	23.1	25.9	28.8	29.3	31.9	31.5	38.7	36.2
22 Madagascar	51.1	46.5	41.9	37.3	35.9	40.0	60.8	60.3	68.6	64.4
23 Malawi	36.4	38.1	37.2	34.7	39.7	34.9	44.8	56.0	97.1	100.2
24 Mali	76.9	63.0	55.0	52.8	59.9	61.1	89.0	87.1	91.6	85.8
25 Mauritanie	28.8	37.7	33.7	32.6	31.3	35.9	43.8	47.3	47.8	46.4
26 Mauritius	10.6	9.1	8.3	8.2	8.1	8.8	11.6	23.4	16.8	18.5
27 Mozambique	37.8	46.3	49.8	43.8	47.1	50.7	69.9	73.1	90.0	89.2
28 Niger	62.1	59.2	68.3	60.8	61.1	75.6	92.0	94.1	98.6	86.3
29 Nigeria	47.5	47.0	43.1	45.0	38.8	40.2	47.6	54.1	59.8	70.3
30 République Centrafr.	34.1	33.6	31.2	28.9	38.5	34.0	45.8	49.4	49.3	49.3
31 Rwanda	54.5	53.2	49.7	53.8	50.1	60.9	66.4	83.3	91.8	82.5
32 Sao Tomé and Principe										
33 Sénégal	122.5	117.7	102.9	99.5	114.9	92.2	121.7	131.5	129.7	131.5
34 Seychelles	8	6.9	7.2	7.6	6.4	4.8	7.7	9.6	9.5	9.1
35 Sierra Leone	21.4	21.7	21.3	19.2	19.4	22.2	29.6	28.5	30.9	28.9
36 Somalia	92.9	103.2	92.0	113.5	107.4	131.6	159.7	146.5	115.5	117.9
37 Sudan	102.4	131.5	118.0	127.9	121.8	203.8	161.1	147.7	152.1	183.3
38 Swaziland	22.5	21.0	20.7	18.2	15.4	17.1	21.6	21.3	28.3	26.4
39 Tanzania U.Rep. of	172.6	176.4	181.2	173.9	138.7	141.4	163.4	186.7	208.4	199.1
40 Tchad	11.9	16.8	15.3	21.7	18.8	42.6	46.0	52.0	53.6	50.1
41 Togo	28.9	30.4	30.3	27.5	29.9	30.0	42.3	44.5	48.3	46.3
42 Uganda	21	33.8	29.5	34.0	32.2	36.7	49.9	57.0	74.3	87.6
43 Zaïre	167.8	151.6	133.7	114.7	97.0	111.3	140.0	175.6	203.8	150.2
44 Zambia	86.8	76.7	74.7	73.8	67.7	74.8	100.8	102.7	116.6	124.3
45 Zimbabwe	70.2	74.9	42.5	51.2	46.9	53.4	64.5	88.3	91.2	96.9
LDCs	1125	1217	1185	1183	1195	1398	1663	1769	1942	1941
SSA	2181	2179	2032	2042	2002	2210	2745	2963	3244	3224

Source: *Geographical Distribution of Financial Flows to Developing Countries*, OECD, Paris, various years.

ANNEX: Statistics on Technical Cooperation 285

TABLE 3: COUNTRY ALLOCATION OF TECHNICAL COOPERATION GRANTS FROM MAJOR DONORS
1970 and 1989 (in millions of current U.S. dollars)

		FRANCE		GERMANY		U. STATES		U.KINGDOM		BELGIUM	
		1970	1989	1970	1989	1970	1989	1970	1989	1970	1989
1	Angola	..	1.1	0.1	1.2	..	..	..	2.1	..	..
2	Bénin	2.8	9.7	1.0	14.7	..	3.0	0.1	0.4	0.1	0.5
3	Botswana	..	0.7	..	10.2	..	10.0	1.4	10.5	..	..
4	Burkina Faso	5.4	19.6	0.5	14.3	..	4.0	..	0.5	..	0.5
5	Burundi	0.7	9.5	0.4	9.5	..	3.0	..	0.2	6.6	14.4
6	Cameroun	9.5	37.2	1.4	19.7	..	19.0	0.5	4.3	0.3	1.5
7	Cap-Vert	..	2.1	..	..	..	1.0	..	0.1	..	0.1
8	Comores	4.8	7.5	..	0.2	..	1.0	..	0.0	..	1.4
9	Congo	6.5	25.0	0.3	5.9	..	..	..	0.4	..	..
10	Côte d'Iv.	13.2	75.8	1.2	7.5	1.0	1.0	0.1	1.1	0.2	2.7
11	Djibouti	6.2	26.5	..	2.5	..	1.0	..	0.1	..	..
12	Ethiopia	..	7.8	3.7	9.3	5.0	..	1.1	3.6	0.1	0.0
13	Gabon	5.6	22.6	..	0.7	..	1.0	..	0.1	0.1	1.6
14	Gambia	..	0.5	..	2.8	..	6.0	0.4	4.9	..	..
15	Ghana	..	2.9	3.1	12.9	3.0	5.0	2.2	7.4	..	..
16	Guinée	..	9.1	1.2	7.2	..	3.0	..	0.5	..	..
17	Guinée-Bis.	..	2.7	..	0.8	..	2.0	..	0.1	..	..
18	Guinée-Eq.	..	3.8	..	1.5	..	1.0	..	0.0	..	..
19	Kenya	..	2.3	2.2	27.3	4.0	25.0	7.8	26.2	..	1.2
20	Lesotho	..	0.4	..	5.6	..	15.0	0.6	5.2	..	..
21	Liberia	..	0.9	1.1	8.8	5.0	14.0	0.1	1.0	..	..
22	Madagascar	15.1	32.5	1.4	9.1	..	1.0	0.1	0.5	0.1	0.0
23	Malawi	2.1	1.4	..	8.4	1.0	16.0	1.0	15.7	0.1	0.1
24	Mali	4.3	22.0	1.1	15.5	..	17.0	..	1.4	..	0.5
25	Mauritanie	2.6	20.8	0.4	4.6	..	6.0	..	0.2	..	..
26	Mauritius	..	7.7	..	0.6	..	..	0.5	3.4	..	..
27	Mozambique	..	2.7	..	3.9	..	1.0	..	4.3	..	..
28	Niger	6.6	22.7	0.7	11.7	1.0	14.0	..	0.4	0.1	2.6
29	Nigeria	..	5.4	2.0	6.6	17.0	7.0	3.6	10.2	..	0.0
30	Rép Centraf.	5.6	25.7	0.6	5.6	..	5.0	..	0.1	0.1	..
31	Rwanda	0.6	7.2	0.9	18.7	..	7.0	..	0.3	6.4	17.4
32	Sao Tomé	..	..	..	..	..	..	..	..	..	..
33	Sénégal	13.7	57.5	0.5	6.5	1.0	19.0	0.2	1.3	0.4	3.8
34	Seychelles	..	2.9	..	1.3	..	..	0.6	1.8	..	0.4
35	Sierra Leone	..	1.2	0.2	7.8	2.0	2.0	1.0	3.3	..	..
36	Somalia	..	1.2	1.7	17.2	2.0	17.0	0.3	5.4	0.0	0.0
37	Sudan	..	3.9	0.6	20.5	0.0	33.0	1.1	13.8	0.0	0.6
38	Swaziland	..	0.2	..	2.6	..	9.0	1.1	4.3	..	0.4
39	Tanzania	..	1.2	4.1	25.5	1.0	4.0	4.3	14.4	..	0.2
40	Tchad	1.0	14.7	..	4.8	..	7.0	..	0.2	0.3	..
41	Togo	1.4	15.4	2.4	12.8	1.0	6.0	..	0.5	..	0.1
42	Uganda	..	1.0	1.5	10.5	2.0	14.0	4.9	11.4	..	0.2
43	Zaïre	2.5	14.2	2.1	19.1	1.0	24.0	..	0.6	28.4	54.0
44	Zambia	..	1.3	1.2	14.3	..	6.0	6.1	23.5	..	2.3
45	Zimbabwe	..	1.4	0.1	20.4	..	11.0	0.4	12.2	..	0.2
	LDC	44.1	241.0	21.0	246.1	15.0	199.0	16.2	97.5	13.8	38.6
	SSA	110.2	531.9	37.7	410.6	47.0	341.0	39.5	197.9	43.3	106.7

TABLE 3:—Continued

	NETHERLANDS		JAPAN		SWEDEN		ITALY	
	1970	1989	1970	1989	1970	1989	1970	1989
1 Angola	..	2.1	..	..	..	11.5	..	3.7
2 Bénin	..	4.5	..	..	..	..	..	0.9
3 Botswana	..	3.0	..	0.0	0.2	6.4	..	0.3
4 Burkina Faso	0.1	29.0	..	0.4	..	..	..	0.9
5 Burundi	..	0.9	..	0.7	..	..	..	0.2
6 Cameroun	1.2	4.3	..	0.5	..	0.1	..	2.8
7 Cap-Vert	..	..	..	0.2	..	..	..	0.2
8 Comores	..	0.3	..	0.5	..	..	..	0.0
9 Congo	..	0.4	..	0.1	..	..	..	0.4
10 Côte d'Iv.	0.6	1.7	..	1.1	..	..	..	..
11 Djibouti	..	..	..	0.1	..	..	..	0.3
12 Ethiopia	0.2	5.6	0.2	2.2	3.3	8.4	1.5	38.2
13 Gabon	..	0.4	..	0.4	..	..	..	0.0
14 Gambia	..	1.3	..	0.8	..	0.0	..	0.2
15 Ghana	0.3	4.0	0.2	6.6	..	0.2	0.1	0.3
16 Guinée	..	0.6	..	0.7	..	..	..	2.4
17 Guinée-Bis.	..	2.9	..	0.1	..	6.5	..	0.1
18 Guinée-Eq.	..	0.0	..	0.0	..	..	..	1.1
19 Kenya	1.8	19.4	0.5	2.8	2.2	3.4	0.2	4.7
20 Lesotho	..	0.2	..	0.1	0.1	4.8	..	..
21 Liberia	1.1	1.1	..	2.8	..	..	..	..
22 Madagascar	..	0.5	0.1	2.5	..	0.1	0.2	1.5
23 Malawi	..	1.3	0.1	4.7	..	..	..	..
24 Mali	..	2.9	..	0.6	..	..	..	0.4
25 Mauritanie	..	1.1	..	0.1	..	..	0.1	0.3
26 Mauritius	..	0.1	..	3.1	..	0.6	..	..
27 Mozambique	..	4.0	..	0.5	..	19.1	..	7.0
28 Niger	0.1	3.5	..	4.6	..	..	0.1	..
29 Nigeria	1.2	2.2	0.2	6.3	..	..	0.1	2.2
30 Rép Centraf.	..	0.2	..	0.8	..	..	..	0.2
31 Rwanda	0.3	3.2	..	2.8	..	..	0.1	0.7
32 Sao Tomé	..	..	..	..	..	..	..	..
33 Sénégal	0.1	2.7	..	9.1	..	..	0.1	2.3
34 Seychelles	..	0.2	..	0.3	..	..	..	0.1
35 Sierra Leone	0.1	1.0	..	0.5	0.2	..	..	0.6
36 Somalia	..	0.1	0.0	0.5	..	1.7	3.1	35.5
37 Sudan	0.0	21.9	0.0	4.3	..	0.3	0.2	0.4
38 Swaziland	0.1	1.2	..	0.0	0.1	..	..	1.6
39 Tanzania	1.1	24.5	0.7	3.9	1.9	31.5	0.4	1.2
40 Tchad	..	1.7	..	0.0	..	..	..	1.4
41 Togo	..	0.5	..	0.1	..	..	..	..
42 Uganda	0.2	2.5	0.4	0.4	..	0.3	0.2	1.3
43 Zaïre	0.5	3.1	..	3.0	..	..	0.1	2.3
44 Zambia	0.5	11.2	..	13.4	0.4	11.4	..	1.3
45 Zimbabwe	..	7.2	..	3.0	..	9.1	..	2.8
LDC	2.1	116.7	1.4	39.6	5.7	79.0	5.7	93.8
SSA	9.5	178.5	2.4	114.6	8.4	115.4	6.5	119.8

TABLE 3:—Continued

		UNDP 1989	TOT. BILAT. 1970	1989	MULTILATERAL 1970	1989	TOT.TC 1970	1989
1	Angola	5.0	0.1	22.0	..	25.1	0.1	47.1
2	Bénin	5.9	4.3	34.7	1.7	12.3	6.0	47.0
3	Botswana	5.1	1.7	47.9	0.8	9.8	2.5	57.8
4	Burkina Faso	11.7	6.2	71.1	2.0	23.7	8.2	94.8
5	Burundi	5.4	7.8	38.6	2.7	13.0	10.5	51.6
6	Cameroun	6.3	15.0	96.4	2.2	10.3	17.2	106.7
7	Cap-Vert	4.2	0.0	7.7	..	6.4	..	17.1
8	Comores	2.5	4.8	11.0	..	6.1	4.8	17.1
9	Congo	2.2	6.8	32.3	3.3	4.5	10.1	36.8
10	Côte d'Iv.	3.8	17.4	92.2	3.3	7.2	20.7	99.4
11	Djibouti	2.3	6.2	30.6	..	3.3	6.2	33.9
12	Ethiopia	22.6	15.2	81.0	4.4	137.1	19.6	218.1
13	Gabon	2.0	5.9	29.2	0.9	2.4	6.8	31.6
14	Gambia	5.0	0.4	16.8	0.2	6.7	0.6	23.5
15	Ghana	7.2	11.3	43.9	3.6	15.5	14.9	59.4
16	Guinée	10.8	1.2	24.4	3.0	18.8	4.2	43.2
17	Guinée-Bis.	5.3	0.0	15.1	..	10.1	..	25.2
18	Guinée-Eq.	1.8	0.0	7.4	..	4.6	..	12.0
19	Kenya	5.5	22.8	156.4	4.3	18.3	27.1	174.7
20	Lesotho	3.0	0.9	36.5	1.6	10.2	2.5	46.6
21	Liberia	2.8	6.2	28.7	1.8	7.5	8.0	36.2
22	Madagascar	7.4	17.5	48.7	3.3	15.7	20.8	64.4
23	Malawi	10.8	7.6	50.7	1.3	49.5	8.9	100.2
24	Mali	11.8	5.5	62.6	2.6	23.2	8.1	85.8
25	Mauritanie	8.6	3.1	33.3	1.3	13.3	4.4	46.4
26	Mauritius	1.1	0.5	16.4	0.9	2.2	1.4	18.5
27	Mozambique	12.3	0.0	54.1	..	35.1	..	89.2
28	Niger	13.7	8.9	63.6	2.3	22.7	11.2	86.3
29	Nigeria	6.2	26.7	42.2	9.7	28.2	36.4	70.3
30	Rép Centraf.	7.4	6.4	37.6	1.8	11.7	8.2	49.3
31	Rwanda	10.2	9.2	63.2	2.6	19.3	11.8	82.5
32	Sao Tomé	1.1	..		..		..	
33	Sénégal	10.3	17.6	110.9	3.1	20.6	20.7	131.5
34	Seychelles	0.3	0.6	7.9	..	1.2	0.6	9.1
35	Sierra Leone	6.5	3.5	16.7	1.8	12.2	5.3	28.9
36	Somalia	9.6	7.1	81.3	3.4	36.7	10.5	117.9
37	Sudan	5.2	1.9	103.6	4.5	79.7	6.4	183.3
38	Swaziland	1.5	1.3	21.4	0.6	5.0	1.9	26.4
39	Tanzania	6.7	17.6	156.5	3.7	42.6	21.3	199.1
40	Tchad	15.5	9.0	29.8	1.0	20.3	10.0	50.1
41	Togo	12.6	5.2	35.7	2.2	10.6	7.4	46.3
42	Uganda	10.5	10.9	43.4	3.5	44.2	14.4	87.6
43	Zaïre	9.6	37.0	124.4	4.9	25.8	41.9	150.2
44	Zambia	3.6	10.1	103.3	3.1	20.9	13.2	124.3
45	Zimbabwe	3.6	0.6	81.7	0.1	15.2	0.7	96.9
	LDC	228.0	144.6	1255	48.4	683.2	193.0	1940.8
	SSA	306.4	342.0	2313	93.5	908.8	435.5	3224.3

Source: *Geographical Distribution of Financial Flows to Developing Countries.* OECD, Paris, several years; and UNDP data.

TABLE 4: COUNTRY ALLOCATION OF TECHNICAL COOPERATION GRANTS FROM MAJOR DONORS IN PERCENTAGES, 1970 AND 1989

COUNTRY	FRANCE 1970	FRANCE 1989	GERMANY 1970	GERMANY 1989	U.S. 1970	U.S. 1989	U.K. 1970	U.K. 1989	BELGIUM 1970	BELGIUM 1989
1 Angola		0.2	0.3	0.3				1.1		
2 Bénin	2.5	1.8	2.7	3.6		0.9	0.3	0.2	0.2	0.5
3 Botswana		0.1		2.5		2.9	3.5	5.3		0.0
4 Burkina Faso	4.9	3.7	1.3	3.5		1.2		0.3		0.5
5 Burundi	0.6	1.8	1.1	2.3		0.9		0.1	15.2	13.5
6 Cameroun	8.6	7.0	3.7	4.8		5.6	1.3	2.2	0.7	1.4
7 Cap-Vert		0.4		0.0		0.3		0.1		0.1
8 Comores	4.4	1.4		0.0		0.3		0.0		1.3
9 Congo	5.9	4.7	0.8	1.4		0.0		0.2		0.0
10 Côte d'Iv.	12.0	14.3	3.2	1.8	2.1	0.3	0.3	0.6	0.5	2.5
11 Djibouti	5.6	5.0		0.6		0.3		0.1		0.0
12 Ethiopia		1.5	9.8	2.3	0.6	0.0	2.8	1.8	0.2	0.0
13 Gabon	5.1	4.2		0.2		0.3		0.1	0.2	1.5
14 Gambia		0.1		0.7		1.8	1.0	2.5		0.0
15 Ghana		0.5	8.2	3.1	6.4	1.5	5.6	3.7		0.0
16 Guinée		1.7	3.2	1.8		0.9		0.3		0.0
17 Guinée-Bis.		0.7		0.4		0.3		0.0		0.0
18 Guinée-Eq.		0.5		0.2		0.6		0.1		0.0
19 Kenya		0.4	5.8	6.6	8.5	7.3	19.7	13.2		1.1
20 Lesotho		0.1		1.4		4.4	1.5	2.6		0.0
21 Liberia		0.2	2.9	2.1	10.6	4.1	0.3	0.5		0.0
22 Madagascar	13.7	6.1	3.7	2.2		0.3	0.3	0.3	0.2	0.0
23 Malawi	1.9	0.3		2.0	2.1	4.7	2.5	7.9	0.2	0.1
24 Mali	3.9	4.1	2.9	3.8		5.0		0.7		0.5
25 Mauritanie	2.4	3.9	1.1	1.1		1.8		0.1		0.0
26 Mauritius		1.4		0.1		0.0	1.3	1.7		0.0
27 Mozambique		0.5		0.9		0.3		2.2		0.0
28 Niger	6.0	4.3	1.9	2.8	2.1	4.1		0.2	0.2	2.4
29 Nigeria		1.0	5.3	1.6	36.2	2.1	9.1	5.2		0.0
30 Rép. Centrafr.	5.1	4.8	1.6	1.4		1.5		0.1	0.2	0.0
31 Rwanda	0.5	1.4	2.4	4.6		2.1		0.2	14.8	16.3
32 Sao Tomé										
33 Sénégal	12.4	10.8	1.3	1.6	2.1	5.6	0.5	0.7	0.9	3.6
34 Seychelles		0.5		0.3			1.5	0.9		0.4
35 Sierra Leone		0.2	0.5	1.9	4.3	0.6	2.5	1.7		0.0
36 Somalia		0.2	4.5	4.2	4.3	5.0	0.8	2.7	0.0	0.0
37 Sudan		0.7	1.6	5.0	0.0	9.7	2.8	7.0	0.0	0.6
38 Swaziland				0.6		2.6	2.8	2.2		0.4
39 Tanzania		0.2	10.9	6.2	2.1	1.2	10.9	7.3		0.2
40 Tchad	0.9	2.8		1.2		2.1		0.1	0.7	0.0
41 Togo	1.3	2.9	6.4	3.1	2.1	1.8		0.3		0.1
42 Uganda		0.2	3.2	2.6		4.1	15.4	5.8		0.2
43 Zaïre	2.3	2.7	5.6	4.7	2.1	7.0		0.3	65.6	50.6
44 Zambia		0.2	4.0	3.5	4.3	1.8	12.4	11.9		2.2
45 Zimbabwe		0.3	0.3	5.0		3.2	1.0	6.2		0.2
TOTAL %	100	100	100	100	100	100	100	100	100	100
Of which LDC	40	45	56	60	32	58	41	49	32	36
SSA	110	532	38	410	47	341	40	138	43	107

TABLE 4:—Continued

COUNTRY	NETHERLANDS 1970	NETHERLANDS 1989	JAPAN 1970	JAPAN 1989	SWEEDEN 1970	SWEEDEN 1989	ITALY 1970	ITALY 1989
1 Angola		1.2		0.0		10.0		3.1
2 Bénin		2.5		0.0		0.0		0.8
3 Botswana		1.7		0.0	2.4	5.5		0.3
4 Burkina Faso	1.2	16.2		0.3		0.0		0.8
5 Burundi		0.5		0.6		0.0		0.2
6 Cameroun	14.3	2.4		0.4		0.1		2.3
7 Cap-Vert		0.0		0.2		0.0		0.2
8 Comores		0.2		0.4		0.0		0.0
9 Congo		0.2		0.1		0.0		0.3
10 Côte d'Ivoire	7.1	1.0		1.0		0.0		0.0
11 Djibouti		0.0		0.1		0.0		0.3
12 Ethiopia	2.4	3.1	8.3	1.9	39.3	7.3	23.1	31.9
13 Gabon		0.2		0.3		0.0		0.0
14 Gambia		0.7		0.7		0.0		0.2
15 Ghana	3.6	2.2	8.3	5.8		0.2	1.5	0.3
16 Guinée		0.3		0.6		0.0		2.0
17 Guinée-Bis.		0.0		0.0		0.0		0.9
18 Guinée-Eq.		1.6		0.1		5.6		0.1
19 Kenya	21.4	10.9	20.8	19.9	26.2	2.9	3.1	3.9
20 Lesotho		0.1		0.1	1.2	4.2		0.0
21 Liberia		0.6		2.4		0.0		0.0
22 Madagascar		0.3	4.2	2.2		0.1	3.1	1.3
23 Malawi		0.7	4.2	4.1		0.0		0.0
24 Mali		1.6		0.5		0.0		0.3
25 Mauritanie		0.6		0.1		0.0	1.5	0.3
26 Mauritius		0.1		2.7		0.5		0.0
27 Mozambique		2.2		0.4		16.6		5.8
28 Niger	1.2	2.0		4.0		0.0	1.5	0.0
29 Nigeria	14.3	1.2	8.3	5.5		0.0	1.5	1.8
30 Rép Centrafr.		0.1		0.7		0.0		0.2
31 Rwanda	3.6	1.8		2.4		0.0	1.5	0.6
32 Sao Tomé								
33 Sénégal	1.2	1.5		7.9		0.0	1.5	1.9
34 Seychelles		0.1		0.3				0.1
35 Sierra Leone	1.2	0.6		0.4	2.4			0.5
36 Somalia		0.1	0.0	0.4		1.5	47.7	29.6
37 Sudan	0.0	12.3	0.0	3.8		0.3	3.1	0.3
38 Swaziland	1.2	0.7		0.0	1.2	0.0		1.3
39 Tanzania	13.1	13.7	29.2	12.1	22.6	27.3	6.2	1.0
40 Tchad		1.0		0.0		0.0		1.2
41 Togo		0.3		0.1		0.0		0.0
42 Uganda	6.0	1.4		0.3	4.8	0.3		1.1
43 Zaïre	6.0	1.7		2.6		0.0	1.5	1.9
44 Zambia	2.4	6.3	16.7	11.7		9.9	3.1	1.1
45 Zimbabwe		4.0		2.6		7.9		2.3
TOTAL %	100	100	100	100	100	100	100	100
Of which LDC	25	65	58	35	68	69	88	78
SSA	8	179	2	115	8	115	7	

TABLE 4:—Continued

COUNTRY	OTHER BILATERAL 1970	OTHER BILATERAL 1989	TOTAL BILATERAL 1970	TOTAL BILATERAL 1989	UNDP 1989	TOTAL MULTILAT. 1970	TOTAL MULTILAT. 1989	TOTAL 1970	TOTAL 1989
1 Angola	0.0	2.4	0.0	1.0	1.6		2.8	0.0	1.5
2 Bénin	0.8	1.0	1.3	1.5	1.9	1.8	1.4	1.4	1.5
3 Botswana	0.3	4.1	0.5	2.1	1.6	0.9	1.1	0.6	1.8
4 Burkina Faso	0.5	2.2	1.8	3.1	3.7	2.1	2.6	1.9	2.9
5 Burundi	0.3	0.6	2.3	1.7	1.7	2.9	1.4	2.4	1.6
6 Cameroun	5.4	3.6	4.4	4.2	2.0	2.4	1.1	3.9	3.3
7 Cap-Vert	0.0	0.8	0.0	0.3	1.3		0.7		0.5
8 Comores	0.0	0.0	1.4	0.5	0.8		0.7	1.1	0.5
9 Congo	0.0	0.1	2.0	1.4	0.7	3.5	0.5	2.3	1.1
10 Côte d'Iv.	2.8	0.8	5.1	4.0	1.2	3.5	0.8	4.8	3.1
11 Djibouti	0.0	0.0	1.8	1.3	0.7		0.4	1.4	1.1
12 Ethiopia	0.3	2.7	4.4	3.5	7.2	4.7	15.1	4.5	6.8
13 Gabon	0.5	1.0	1.7	1.3	0.6	1.0	0.3	1.6	1.0
14 Gambia	0.0	0.5	0.1	0.7	1.6	0.2	0.7	0.1	0.7
15 Ghana	6.2	2.1	3.3	1.9	2.3	3.9	1.7	3.4	1.8
16 Guinée	0.0	0.4	0.4	1.1	3.5	3.2	2.1	1.0	1.3
17 Guinée-Bis.	0.0	0.9	0.0	0.3	0.6		0.5		0.4
18 Guinée-Eq.	0.0	0.0	0.0	0.7	1.7		1.1		0.8
19 Kenya	10.6	11.6	6.7	6.8	1.8	4.6	2.0	6.2	5.4
20 Lesotho	0.5	3.3	0.3	1.6	1.0	1.7	1.1	0.6	1.4
21 Liberia	0.0	0.3	1.8	1.2	0.9	1.9	0.8	1.8	1.1
22 Madagascar	1.3	1.3	5.1	2.1	2.4	3.5	1.7	4.8	2.0
23 Malawi	8.5	1.7	2.2	2.2	3.5	1.4	5.4	2.0	3.1
24 Mali	0.3	2.0	1.6	2.7	3.8	2.8	2.6	1.9	2.7
25 Mauritanie	0.0	0.4	0.9	1.4	2.8	1.4	1.5	1.0	1.4
26 Mauritius	0.0	0.4	0.1	0.7	0.4	1.0	0.2	0.3	0.6
27 Mozambique	0.0	4.7	0.0	2.3	4.0		3.9		2.8
28 Niger	0.8	2.9	2.6	2.7	4.3	2.5	2.5	2.6	2.7
29 Nigeria	6.7	0.7	7.8	1.8	2.0	10.4	3.1	8.4	2.2
30 Rép. Centrafr.	0.3	0.1	1.9	1.6	2.4	1.9	1.3	1.9	1.5
31 Rwanda	2.3	3.8	2.7	2.7	3.3	2.8	2.1	2.7	2.6
32 Sao Tomé									
33 Sénégal	4.1	3.1	5.1	4.8	3.2	3.3	2.3	4.8	4.1
34 Seychelles	0.0	0.6	0.2	0.3	0.1		0.1	0.1	0.3
35 Sierra Leone	0.0	0.2	1.0	0.7	2.1	1.9	1.3	1.2	0.9
36 Somalia	0.0	0.5	2.1	3.5	3.1	3.6	4.0	2.4	3.7
37 Sudan	0.0	2.3	0.6	4.5	1.7	4.8	8.8	1.5	5.7
38 Swaziland	0.0	0.8	0.4	0.9	0.5	0.6	0.6	0.4	0.8
39 Tanzania	10.6	17.3	5.1	6.8	3.4	4.0	4.7	4.9	6.2
40 Tchad	19.9	0.7	2.6	1.3	5.0	1.1	2.2	2.3	1.6
41 Togo	1.0	0.3	1.5	1.5	2.2	2.4	1.2	1.7	1.4
42 Uganda	4.9	9.6	3.0	1.9	4.1	3.3	4.9	3.0	2.7
43 Zaïre	6.2	2.3	10.8	5.4	3.1	5.2	2.8	9.6	4.7
44 Zambia	4.4	0.8	3.2	4.5	1.2	3.7	2.3	3.3	3.9
45 Zimbabwe	0.3	5.1	0.2	3.5	1.2	0.1	1.7	0.2	3.0
TOTAL %	100	100	100	100	100	100	100	100	100
Of which LDC	51	54	42	54	73	52	75	44	60
SSA	39	220	342	2313	310	94	909	436	3224

Source: Based on OECD data for DAC Countries and multilateral total, and internal data for UNDP.

TABLE 5: TECHNICAL COOPERATION GRANTS TO SUB-SAHARAN AFRICA: DISTRIBUTION BY MAJOR DONOR SOURCES IN PERCENTAGES, 1970 AND 1989

		FRANCE		GERMANY		U. STATES		U.KINGDOM		BELGIUM	
		1970	1989	1970	1989	1970	1989	1970	1989	1970	1989
1	Angola	0.0	2.3		2.5	0.0	0.0	0.0	4.5	0.0	0.0
2	Bénin	46.7	20.6	16.7	31.3	0.0	6.4	1.7	0.9	1.7	1.1
3	Botswana	0.0	1.2	0.0	17.7	0.0	17.3	56.0	18.2	0.0	0.0
4	Burkina Faso	65.9	20.7	6.1	15.1	0.0	4.2	0.0	0.5	0.0	0.5
5	Burundi	6.7	18.4	3.8	18.4	0.0	5.8	0.0	0.4	62.9	27.9
6	Cameroun	55.2	34.9	8.1	18.5	0.0	17.8	2.9	4.0	1.7	1.4
7	Cap-Vert		12.3		0.0		5.8		0.6		0.6
8	Comores	100	43.9	0.0	1.2	0.0	5.8	0.0	0.0	0.0	8.2
9	Congo	64.4	67.9	3.0	16.0	0.0	0.0	0.0	1.1	0.0	0.0
10	Côte d'Ivoire	63.8	76.3	5.8	7.5	4.8	1.0	0.5	1.1	1.0	2.7
11	Djibouti	100	78.2	0.0	7.4	0.0	2.9	0.0	0.3	0.0	0.0
12	Ethiopia	0.0	3.6	18.9	4.3	25.5	0.0	5.6	1.7	0.5	0.0
13	Gabon	82.4	71.5	0.0	2.2	0.0	3.2	0.0	0.3	1.5	5.1
14	Gambia	0.0	2.1	0.0	11.9	0.0	25.5	66.7	20.9	0.0	0.0
15	Ghana	0.0	4.9	20.8	21.7	20.1	8.4	14.8	12.5	0.0	0.0
16	Guinée	0.0	21.1	28.6	16.7	0.0	6.9	0.0	1.2	0.0	0.0
17	Guinée-Bis.		10.7		3.2		7.9		0.4		0.0
18	Guinée-Eq.		31.7		12.5		8.3		0.0		0.0
19	Kenya	0.0	1.3	8.1	15.6	14.8	14.3	28.8	15.0	0.0	0.7
20	Lesotho	0.0	0.9	0.0	12.0	0.0	32.2	24.0	11.2	0.0	0.0
21	Liberia	0.0	2.5	13.8	24.3	62.5	38.7	1.3	2.8	0.0	0.0
22	Madagascar	72.6	50.5	6.7	14.1	0.0	1.6	0.5	0.8	0.5	0.0
23	Malawi	23.6	1.4	0.0	8.4	11.2	16.0	11.2	15.7	1.1	0.1
24	Mali	53.1	25.6	13.6	18.1	0.0	19.8	0.0	1.6	0.0	0.6
25	Mauritanie	59.1	44.8	9.1	9.9	0.0	12.9	0.0	0.4	0.0	0.0
26	Mauritius	0.0	41.6	0.0	3.2	0.0	0.0	35.7	18.4	0.0	0.0
27	Mozambique		3.0		4.4		1.1		4.8		0.0
28	Niger	58.9	26.3	6.3	13.6	8.9	16.2	0.0	0.5	0.9	3.0
29	Nigeria	0.0	7.7	5.5	9.4	46.7	10.0	9.9	14.5	0.0	0.0
30	République Centrafr.	68.3	52.1	7.3	11.4	0.0	10.1	0.0	0.2	1.2	0.0
31	Rwanda	5.1	8.7	7.6	22.7	0.0	8.5	0.0	0.4	54.2	21.1
32	Sao Tomé										
33	Sénégal	66.2	43.7	2.4	4.9	4.8	14.4	1.0	1.0	1.9	2.9
34	Seychelles	0.0	31.9	0.0	14.3	0.0	0.0		19.8	0.0	4.4
35	Sierra Leone	0.0	4.2	3.8	27.0	37.7	6.9	18.9	11.4	0.0	0.0
36	Somalia	0.0	1.0	16.2	14.6	19.0	14.4	2.9	4.6	0.0	0.0
37	Sudan	0.0	2.1	9.4	11.2	0.0	18.0	17.2	7.5	0.0	0.3
38	Swaziland	0.0	0.8	0.0	9.8	0.0	34.1	57.9	16.3	0.0	1.5
39	Tanzania	0.0	0.6	19.2	12.8	4.7	2.0	20.2	7.2	0.0	0.1
40	Tchad	10.0	29.3	0.0	9.6	0.0	14.0	0.0	0.4	3.0	0.0
41	Togo	18.9	33.3	32.4	27.6	13.5	13.0	0.0	1.1	0.0	0.2
42	Uganda	0.0	1.1	10.4	12.0	13.9	16.0	34.0	13.0	0.0	0.2
43	Zaïre	6.0	9.5	5.0	12.7	2.4	16.0	0.0	0.4	67.8	36.0
44	Zambia	0.0	1.0	9.1	11.5	0.0	4.8	46.2	18.9	0.0	1.9
45	Zimbabwe	0.0	1.4	14.3	21.1	0.0	11.4	57.1	12.6	0.0	0.2
	LDC	22.8	12.4	10.9	12.7	7.8	10.3	8.4	5.0	7.2	2.0
	SSA	25.3	16.5	8.7	12.7	10.8	10.6	9.1	6.1	9.9	3.3

TABLE 5:—Continued

	NETHERLANDS		JAPAN		SWEDEN		ITALY	
	1970	1989	1970	1989	1970	1989	1970	1989
1 Angola	0.0	4.5	0.0	0.0	0.0	24.4	0.0	7.9
2 Bénin	0.0	9.6	0.0	0.0	0.0	0.0	0.0	1.9
3 Botswana	0.0	5.2	0.0	0.0	8.0	11.1	0.0	0.5
4 Burkina Faso	1.2	30.6	0.0	0.4	0.0	0.0	0.0	0.9
5 Burundi	0.0	1.7	0.0	1.4	0.0	0.0	0.0	0.4
6 Cameroun	7.0	4.0	0.0	0.5	0.0	0.1	0.0	2.6
7 Cap-Vert		0.0		1.2		0.0		1.2
8 Comores	0.0	1.8	0.0	2.9	0.0	0.0	0.0	0.0
9 Congo	0.0	1.1	0.0	0.3	0.0	0.0	0.0	1.1
10 Côte d'Ivoire	2.9	1.7	0.0	1.1	0.0	0.0	0.0	0.0
11 Djibouti	0.0	0.0	0.0	0.3	0.0	0.0	0.0	0.9
12 Ethiopia	1.0	2.6	1.0	1.0	16.8	3.9	7.7	17.5
13 Gabon	0.0	1.3	0.0	1.3	0.0	0.0	0.0	0.0
14 Gambia	0.0	5.5	0.0	3.4	0.0	0.0	0.0	0.9
15 Ghana	2.0	6.7	1.3	11.1	0.0	0.3	0.7	0.5
16 Guinée	0.0	1.4	0.0	1.6	0.0	0.0	0.0	5.6
17 Guinée-Bissau		11.5		0.4		25.8		0.4
18 Guinée-Equitoriale		0.0		0.0		0.0		9.2
19 Kenya	6.6	11.1	1.8	13.1	8.1	1.9	0.7	2.7
20 Lesotho	0.0	0.4	0.0	0.2	4.0	10.3	0.0	0.0
21 Liberia	13.8	3.0	0.0	7.7	0.0	0.0	0.0	0.0
22 Madagascar	0.0	0.8	0.5	3.9	0.0	0.2	1.0	2.3
23 Malawi	0.0	1.3	1.1	4.7	0.0	0.0	0.0	0.0
24 Mali	0.0	3.4	0.0	0.7	0.0	0.0	0.0	0.5
25 Mauritanie	0.0	2.4	0.0	0.2	0.0	0.0	2.3	0.6
26 Mauritius	0.0	0.5	0.0	16.8	0.0	3.2	0.0	0.0
27 Mozambique		4.5		0.6		21.4		7.8
28 Niger	0.9	4.1	0.0	5.3	0.0	0.0	0.9	0.0
29 Nigeria	3.3	3.1	0.5	9.0	0.0	0.0	0.3	3.1
30 République Centrafr.	0.0	0.4	0.0	1.6	0.0	0.0	0.0	0.4
31 Rwanda	2.5	3.9	0.0	3.4	0.0	0.0	0.8	0.8
32 Sao Tomé and Principe								
33 Sénégal	0.5	2.1	0.0	6.9	0.0	0.0	0.5	1.7
34 Seychelles	0.0	2.2	0.0	3.3	0.0	0.0	0.0	1.1
35 Sierra Leone	1.9	3.5	0.0	1.7	3.8	0.0	0.0	2.1
36 Somalia	0.0	0.1	0.0	0.4	0.0	1.4	29.5	30.1
37 Sudan	0.0	11.9	0.0	2.3	0.0	0.2	3.1	0.2
38 Swaziland	5.3	4.5	0.0	0.0	5.3	0.0	0.0	6.1
39 Tanzania	5.2	12.3	3.3	7.0	8.9	15.8	1.9	0.6
40 Tchad	0.0	3.4	0.0	0.0	0.0	0.0	0.0	2.8
41 Togo	0.0	1.1	0.0	0.2	0.0	0.0	0.0	0.0
42 Uganda	1.4	2.9	2.8	0.5	0.0	0.3	1.4	1.5
43 Zaïre	1.2	2.1	0.0	2.0	0.0	0.0	0.2	1.5
44 Zambia	3.8	9.0	0.0	10.8	3.0	9.2	0.0	1.0
45 Zimbabwe	0.0	7.4	0.0	3.1	0.0	9.4	0.0	2.9
LDC	1.1	6.0	0.7	2.0	3.0	4.1	3.0	4.8
SSA	2.2	5.5	0.6	3.6	1.9	3.6	1.5	3.7

TABLE 5:—Continued

	TOT. BILAT. 1970	TOT. BILAT. 1989	UNDP 1989	MULTILATERAL 1970	MULTILATERAL 1989	TOT. TC 1970	TOT. TC 1989
1 Angola	100	47	10	0	53	0.1	47
2 Bénin	72	74	13	28	26	6.0	47
3 Botswana	68	83	9.0	32	17	2.5	58
4 Burkina Faso	76	75	12	24	25	8.2	95
5 Burundi	74	75	11	26	25	11	52
6 Cameroun	87	90	26	13	9.7	17	107
7 Cap-Vert		45	25		37		17
8 Comores	100	64	15	0	36	4.8	17
9 Congo	67	88	6.2	33	12	10	37
10 Côte d'Ivoire	84	93	3.9	16	7.2	21	99
11 Djibouti	100	90	6.8	0	9.7	6.2	34
12 Ethiopia	78	37	10	22	63	20	218
13 Gabon	87	92	6.3	13	7.6	6.8	32
14 Gambia	67	71	21	33	29	0.6	24
15 Ghana	76	74	12	24	26	15	59
16 Guinée	29	56	25	71	44	4.2	43
17 Guinée-Bissau		60	21		40		25
18 Guinée-Equatoriale		62	16		38		12
19 Kenya	84	90	3.2	16	10	27	175
20 Lesotho	36	78	6.6	64	22	2.5	47
21 Liberia	78	79	8.0	23	21	8.0	36
22 Madagascar	84	76	12	16	24	21	64
23 Malawi	85	51	11	15	49	8.9	100
24 Mali	68	73	14	32	27	8.1	86
25 Mauritanie	70	72	19	30	29	4.4	46
26 Mauritius	36	89	5.9	64	12	1.4	19
27 Mozambique		61	14		39		89
28 Niger	79	74	16	21	26	11	86
29 Nigeria	73	60	9	27	40	36	70
30 République Centrafr.	78	76	15	22	24	8.2	49
31 Rwanda	78	77	12	22	23	12	83
32 Sao Tomé and Principe							
33 Sénégal	85	84	7.6	15	16	21	132
34 Seychelles	100	87	3.3	0	13	0.6	9.1
35 Sierra Leone	66	58	23	34	42	5.3	29
36 Somalia	68	69	8.2	32	31	11	118
37 Sudan	30	57	2.8	70	43	6.4	183
38 Swaziland	68	81	5.7	32	19	1.9	26
39 Tanzania	83	79	5.3	17	21	21	199
40 Tchad	90	59	31	10	41	10	50
41 Togo	70	77	15	30	23	7.4	46
42 Uganda	76	50	15	24	50	14	88
43 Zaïre	88	83	6.5	12	17	42	150
44 Zambia	77	83	2.9	23	17	13	124
45 Zimbabwe	86	84	3.7	14	16	0.7	97
LDC	75	65	12	25	35	193	1941
SSA	79	72	10	21	28	436	3224

Source: Based on OECD data for DAC Countries and multilateral donors; total and internal data for UNDP.

TABLE 6: DISTRIBUTION OF COOPERATION GRANTS BY PRINCIPAL ECONOMIC SECTORS IN 1988
(percentage)

Country	Natural Resources %	Agriculture, Forests and Fisheries %	Industry and Trade %	Transportation and Communication %	Health %	Education %	Others %
1 Angola	5.9	36.8	3.3	9.2	25.0	9.5	10.3
2 Bénin	9.0	17.0	3.0	7.0	15.0	31.0	18.0
3 Botswana	11.0	11.6	3.7	6.4	9.0	23.4	34.9
4 Burkina Faso	6.0	50.0	1.0	2.0	12.0	5.0	24.0
5 Burundi	6.4	23.2	0.8	7.2	11.5	26.7	24.2
6 Cameroun	0.5	21.5	2.0	6.1	16.8	40.7	12.5
7 Cap-Vert	6.1	43.0	1.0	6.3	9.9	19.9	13.6
8 Comores	4.7	7.8	1.9	12.5	18.1	41.9	13.0
9 Congo	2.1	6.1	1.7	11.7	4.5	6.7	67.2
10 Côte d'Ivoire	2.4	9.7	7.8	1.0	3.1	37.7	38.4
11 Djibouti	7.0	7.3	1.7	4.1	9.4	45.2	25.6
12 Ethiopia	-	-	-	-	-	-	-
13 Gabon	-	-	-	-	-	-	-
14 Gambia	14.0	26.0	4.0	9.0	12.0	14.0	21.0
15 Ghana	10.2	13.5	17.6	2.6	9.9	19.1	27.1
16 Guinée	6.2	17.4	3.7	13.7	6.3	23.4	29.3
17 Guinée-Bissau	10.0	31.0	9.0	3.0	6.0	9.0	32.0
18 Guinée-Equatoriale	6.7	15.4	0.1	5.1	21.2	19.9	31.7
19 Kenya	-	-	-	-	-	-	-
20 Lesotho	16.9	27.6	2.3	4.6	6.4	30.5	11.7
21 Liberia	5.4	16.9	3.9	9.0	10.3	12.0	42.5
22 Madagascar	2.0	23.4	2.6	7.6	9.6	48.9	5.9
23 Malawi	4.0	30.0	10.0	16.0	13.0	12.0	14.0
24 Mali	10.0	24.7	3.6	6.4	8.8	26.9	19.6
25 Mauritanie	36.0	23.0	7.0	11.0	5.0	1.0	18.0
26 Mauritius	-	-	-	-	-	-	-
27 Mozambique	4.8	23.4	10.2	21.2	9.0	17.2	14.3
28 Niger	19.3	39.6	4.1	2.0	7.0	14.2	14.0
29 Nigeria	3.7	29.6	5.7	0.5	17.2	11.0	32.4
30 République-Centrafr.	3.7	16.5	0.2	13.4	17.4	26.2	22.8

TABLE 6:—Continued

Country	Natural Resources %	Agriculture, Forests and Fisheries %	Industry and Trade %	Transportation and Communication %	Health %	Education %	Others %
31 Rwanda	15.0	19.0	4.0	16.0	18.0	11.0	17.0
32 Sao Tomé and Principe	0.0	51.0	0.0	9.0	14.0	21.0	5.0
33 Sénégal	9.4	22.2	2.7	0.6	10.1	39.7	15.3
34 Seychelles	3.4	10.9	0.8	1.4	8.7	5.0	69.8
35 Sierra Leone	2.0	18.0	2.0	12.0	28.0	16.0	22.0
36 Somalia	11.9	21.9	6.5	39.0	4.2	0.5	15.9
37 Sudan	-	-	-	-	-	-	-
38 Swaziland	5.4	11.8	3.5	7.4	15.5	42.3	14.1
39 Tanzania	5.9	27.5	10.3	18.4	20.0	6.9	10.8
40 Tchad	6.6	32.7	2.2	13.1	15.4	15.6	14.5
41 Togo	1.6	14.0	7.1	6.1	16.5	17.5	37.3
42 Uganda	8.0	20.6	0.7	8.6	20.0	13.7	27.6
43 Zaïre	6.1	21.2	0.8	13.8	16.6	17.5	23.8
44 Zambia	9.0	42.0	2.0	5.0	9.0	13.0	19.0
45 Zimbabwe	5.4	12.1	4.7	6.4	12.3	25.6	33.2
SSA	7.6	22.9	4.0	8.9	12.5	20.5	23.6
LDC	9.0	24.6	3.8	10.5	12.8	18.8	20.5

Source: Based on the UNDP Development Cooperation Reports.

TABLE 7: ECONOMIC SIGNIFICANCE OF TECHNICAL COOPERATION GRANTS FOR RECIPIENT COUNTRIES

	7.1				7.2			
	GNP Per Capita (US dollars)				Technical Cooperation Per Capita (U.S. $)			
COUNTRY	1970	1980	1985	1989	1970	1980	1985	1989
1 Angola		730	1040			2	3	5
2 Bénin	110	320	260	380	2	7	7	10
3 Botswana	130	790	870		4	59	34	47
4 Burkina Faso	60	180	140	310	2	13	9	11
5 Burundi	70	200	250	220	3	11	10	10
6 Cameroun	180	760	840	1000	3	10	6	9
7 Cap-Vert		380	440	760			58	46
8 Comores	100	340	300	450	20	20	28	37
9 Congo	240	850	1060	940	11	24	17	17
10 Côte d'Ivoire	270	1170	630	790	4	12	4	9
11 Djibouti		970	1200			79	82	83
12 Ethiopia	60	120	110	120	1	1	2	5
13 Gabon	670	3830	3320	2960	14	57	30	29
14 Gambia	100	360	210	240	2	21	24	28
15 Ghana	250	400	370	390	2	4	2	4
16 Guinée	120	280	320	430	1	3	3	8
17 Guinée-Bissau		140	190	180		15	17	26
18 Guinée-Equatoriale	210						16	35
19 Kenya	130	410	310	370	2	8	6	8
20 Lesotho	100	420	420	470	3	22	21	27
21 Liberia	310	590	480		5	13	13	15
22 Madagascar	130	350	240	230	3	6	4	55
23 Malawi	60	180	170	180	2	6	5	12
24 Mali	70	240	150	270	2	11	8	10
25 Mauritanie	180	440	400	490	4	18	21	24
26 Mauritius	280	1190	1100	1950	2	11	8	5
27 Mozambique			170	80		4	4	6
28 Niger	160	430	240	290	3	12	12	12
29 Nigeria	150	1020	950	250	1	1	0	1
30 République Centrafr.	100	320	280	390	5	15	13	17
31 Rwanda	60	240	280	320	3	11	10	12
32 Sao Tomé et Principe	390	380	310					
33 Sénégal	210	490	370	650	5	22	14	18
34 Seychelles	350	2010	2520	4170	12	121	74	134
35 Sierra Leone	160	320	360	200	2	6	6	7
36 Somalia	90	140	160		4	24	24	19
37 Sudan	140	440	320	470	0	6	9	8
38 Swaziland	270	820	740	900	4	40	23	45
39 Tanzania, U.Rep. of	100	270	290	120	2	10	6	8
40 Tchad	100	160	160	190	3	3	9	9
41 Togo	140	430	240	390	4	12	10	13
42 Uganda	190	280	230		1	2	2	5
43 Zaïre	180	430	160	260	2	6	4	4
44 Zambia	450	600	370	390	3	15	11	16
45 Zimbabwe	280	710	590	650		9	6	10
LDC	105	265	231	210	2	7	7	10
SSA	140	548	420	318	2	6	5	7
SSA excl. Nigeria	145	400	316	311	2	8	7	9

TABLE 7:—Continued

COUNTRY	7.3 Technical Cooperation as Percentage of GDP				7.4 Technical Cooperation as Percentage of ODA			
	1970	1980	1985	1989	1970	1980	1985	1989
1 Angola		0.2	0.3	0.7	0	30	28	10
2 Bénin	2.5	2.3	3.0	2.6	40	28	32	19
3 Botswana	3.6	5.7	4.1		18	45	38	36
4 Burkina Faso	2.4	5.1	6.5	3.4	37	34	35	34
5 Burundi	5.0	4.7	4.4	4.4	59	38	32	26
6 Cameroun	1.6	1.3	0.8	0.9	29	32	39	23
7 Cap-Vert			14.7	6.0	-	-	27	23
8 Comores	16.0	5.8	10.1	8.0	61	16	23	38
9 Congo	3.7	2.3	1.5	1.7	62	40	44	41
10 Côte d'Ivoire	1.4	1.1	0.7	1.0	39	47	35	24
11 Djibouti		9.3	7.2		53	39	36	45
12 Ethiopia	1.0	1.1	2.2	3.6	49	21	15	29
13 Gabon	2.2	1.0	0.9	1.0	29	67	49	24
14 Gambia	1.5	5.3	11.9	11.9	46	23	36	25
15 Ghana	0.6	1.0	0.7	1.0	25	22	15	11
16 Guinée	0.9	1.1	1.1	1.8	41	21	17	12
17 Guinée-Bissau		12.1	9.9	14.5	-	20	26	25
18 Guinée-Equatoriale				8.0	-	-	35	29
19 Kenya	1.6	1.9	2.1	1.6	47	32	27	15
20 Lesotho	3.1	4.2	5.5	5.7	25	31	33	36
21 Liberia	2.2	2.2	2.8		62	24	32	61
22 Madagascar	2.4	1.6	1.8	2.5	43	22	21	20
23 Malawi	3.1	3.2	3.1	6.7	24	25	31	25
24 Mali	2.5	4.6	5.6	4.0	38	29	16	19
25 Mauritanie	2.6	4.3	5.5	4.8	59	16	17	19
26 Mauritius	0.7	0.9	0.9	0.9	23	32	31	32
27 Mozambique		1.6	1.5	7.4	-	22	17	12
28 Niger	2.9	2.5	5.3	3.9	35	37	25	29
29 Nigeria	0.5	0.1	0.1	0.2	34	133	124	20
30 République Centrafr.	3.7	4.3	4.7	4.3	57	31	33	26
31 Rwanda	5.9	4.7	3.6	3.8	54	35	34	35
32 Sao Tomé and Principe					-	-	-	
33 Sénégal	2.4	4.3	3.8	2.7	48	47	31	20
34 Seychelles	3.0	5.7	2.8	3.1	15	37	22	46
35 Sierra Leone	1.1	2.0	1.9	3.5	77	24	34	29
36 Somalia	5.5	6.1	10.2	11.3	38	21	37	28
37 Sudan	0.3	1.4	3.0	n.a.	100	17	18	24
38 Swaziland	2.4	3.8	4.2	4.0	31	45	67	96
39 Tanzania, U.Rep. of	1.6	3.4	2.3	6.4	42	25	29	22
40 Tchad	3.3	1.6	6.5	4.8	45	34	23	21
41 Togo	2.7	2.6	3.5	3.3	44	32	26	25
42 Uganda	1.1	0.7	0.7	2.0	44	18	20	22
43 Zaïre	2.6	1.7	2.6	1.7	47	39	34	24
44 Zambia	0.8	2.4	3.5	4.0	99	27	23	32
45 Zimbabwe		1.3	1.1	1.5	100	43	23	36
LDC	1.5	2.7	3.1	4.5	43	25	24	27
SSA	1.2	1.1	1.2	2.0	42	29	26	23
SSA excl. Nigeria	1.4	2.0	2.1	2.5	42	29	25	23

TABLE 7:—Continued

Country	7.5 TC as Percentage of Government Revenue			7.6 Technical Cooperation as Percentage of Exports			
	1980	1985	1987	1970	1980	1985	1970
1 Angola				-	-	-	-
2 Bénin	17	23		10	12	20	24
3 Botswana	14	6		11	9	5	4
4 Burkina Faso	36	40	30	33	45	53	48
5 Burundi	37	26		44	69	41	42
6 Cameroun	8	3	4	7	6	3	7
7 Cap-Vert				-	-	637	-
8 Comores		77	71	96	64	74	-
9 Congo	5	4	12	36	4	3	5
10 Côte d'Ivoire	4	2		4	3	1	3
11 Djibouti				-	-	-	-
12 Ethiopia	6	9	11	16	10	31	45
13 Gabon	2	2	5	5	2	2	3
14 Gambia	24	40	28	4	41	42	53
15 Ghana	14	6		3	3	5	6
16 Guinée	4	6		-	-	-	-
17 Guinée-Bissau	75	81	132	-	110	124	-
18 Guinée-Equitoriale				-	-	-	-
19 Kenya	8	9	9	9	9	12	17
20 Lesotho	22	29	28	42	50	142	87
21 Liberia	12	14	17	4	4	7	10
22 Madagascar	8	11	16	14	13	14	23
23 Malawi	15	14	23	15	13	14	33
24 Mali	43	34	29	23	38	34	35
25 Mauritanie	24	23	22	5	15	10	11
26 Mauritius	5	4	7	2	2	2	1
27 Mozambique				-	13	55	-
28 Niger	17	51	39	35	11	30	39
29 Nigeria	0	0	1	3	0	0	1
30 République Centrafr.		37		19	23	26	38
31 Rwanda	37	28	31	47	41	48	73
32 Sao Tomé et Principe				-	-	-	-
33 Sénégal	17	20	15	13	26	19	18
34 Seychelles	14	7	9	30	38	17	31
35 Sierra Leone	12	24	64	5	10	17	28
36 Somalia	56	100		34	70	145	196
37 Sudan	11	28	16	2	17	34	30
38 Swaziland	12	16	16	3	6	10	-
39 Tanzania, U.Rep. of	19	13	33	9	32	55	52
40 Tchad		90		40	17	48	-
41 Togo	8	13	15	11	6	12	14
42 Uganda	39	15	44	6	6	11	20
43 Zaire	12	14	23	6	7	7	10
44 Zambia	9	13	22	1	6	9	10
45 Zimbabwe	5	4	5	0	5	5	6
LDC	17	19	23	12	19	28	34
SSA	5	7	11	6	4	7	11
SSA excl. Nigeria	10	11	14	7	9	10	15

TABLE 7.7: TECHNICAL COOPERATION GRANTS PER CAPITA, RANK ORDERING, 1989
(in US dollars)

COUNTRY	TC Per Capita	GNP Per Capita
1 Seychelles	133.8	$4,170
2 Djibouti	82.7	-
3 Madagascar	54.9	$230
4 Botswana	47.4	-
5 Cap-Vert	46.3	$760
6 Comores	37.3	$450
7 Guinée-Equatoriale	34.9	-
8 Swaziland	34.7	$900
9 Gabon	28.6	$2,960
10 Gambia	27.7	$240
11 Lesotho	27.1	$470
12 Guinée-Bissau	26.3	$180
13 Mauritania	23.8	$490
14 Somalia	19.4	-
15 Sénégal	18.2	$650
16 République Centrafr.	16.7	$390
17 Congo	16.7	$940
18 Zambia	15.9	$390
19 Liberia	14.6	-
20 Togo	13.2	$390
21 Malawi	12.2	$180
22 Rwanda	12.0	$320
23 Niger	11.5	$290
24 Burkina Faso	10.8	$310
25 Mali	10.4	$270
26 Bénin	10.2	$380
27 Zimbabwe	10.1	$650
28 Burundi	9.7	$220
29 Cameroun	9.2	$1,000
30 Tchad	9.0	$190
31 Côte d'Ivoire	8.5	$790
32 Guinée	7.8	$430
33 Tanzania	7.8	$120
34 Kenya	7.5	$370
35 Sudan	7.5	$470
36 Sierra Leone	7.2	$200
37 Mozambique	5.8	$80
38 Uganda	5.2	-
39 Angola	4.9	-
40 Mauritius	4.6	$1,950
41 Ethiopia	4.5	$120
42 Zaïre	4.4	$260
43 Ghana	4.1	$390
44 Nigeria	0.6	$250
45 Sao Tomé and Principe	-	-

Source: Based on World Bank (GNP, public revenue, export earnings) and OECD data (TC and Population).

BIBLIOGRAPHY

Books and Journal Articles

Adams, Martin E. "Aid Coordination in Africa: A Review." *Development Policy Review*, vol. 7, 1989.

Adediji, A. "Strategy and Tactics for Administrative Reform in Africa." In Ryweyemamu, A., and Hyden, G., eds. *A Decade of Public Administration in Africa*. Nairobi, East Africa Literature Bureau, 1975.

Alexander, Yonah. *International Technical Assistance Experts*. A Case Study of the U.N. Experience. Praeger Special Studies in International Economics and Development, 1966.

Althabe, Gérard. "Le Poids de l'Assistance Technique." *Le Mois en Afrique*, no. 78, juin 1972.

Baldwin, Stephen. "Technical Cooperation: Process, Problems and Prospects." *Journal of World Affairs*, vol. 150, no. 4, Spring 1988: 239-250.

Belshaw, Cyril S. "Evaluation of Technical Assistance as a Contribution to Development." *International Development Review*, 8, no. 2, June 1966: 2-6, 23.

Brent, Stephen. "Aiding Africa." *Foreign Policy*, 80, Fall 1990: 121-140.

Caiden, Gerald. "Administrative Reform: A Prospectus." *International Review of Administrative Sciences*, 44:1-2, May 1978.

_____. *Administrative Reform*. Chicago: Aldine, 1969.

"Changing Roles: SIDA's Revolution." *Development Journal*, February 1990: 8-12.

Conyers, Diana, and Warren, Dennis M. "The Role of Integrated Rural Development Projects in Developing Local Institutional Capacity." *Manchester Papers on Development*, vol. 4, no. 1, 1988: 28-41.

de Lusignan, Guy. "Mieux Rentabiliser l'Assistance Technique par une Meilleure Gestion." *Revue Française d'Administration Publique*, 1989.

Delage, Paul B. "In Search of a Synthesis: From Organizational Theories to the Institutional Building Model for the Analysis of TA Projects." *International Journal of Project Management,* vol. 7, May 1989: 92-100.

Dror, Y. "Strategies for Administrative Reform." In Leemans, A.F., ed. *The Management of Change in Government*, The Hague, 1976.

Esman, Milton J., and Montgomery, John D. "Systems Approaches to Technical Cooperation: The Role of Development Administration." *Public Administration Review*. September/October 1969: 507-539.

France, Gouvernement de. "L'Assistance Technique aux Pays Insuffisamment Développés." *Notes et Etudes Documentaires*, octobre 1954.

Gow, David. "Collaboration in Development Consulting: Stooges, Hired Guns, or Musketeers?" *Human Organization*, vol. 50, no. 1, 1991: 1-14.

_____. "Provision of Technical Assistance: A View From the Trenches." *Canadian Journal of Development Studies,* vol. 9, no. 1, 1988: 81-103.

Guba, Egon, and Lincoln, Yvonna S. *Fourth Generation Evaluation*. Newbury Park, California: Sage Publications, 1989.

Honadle, George; Gow, David; and Silverman, Jerry. "Technical Assistance Alternatives for Rural Development: Beyond the Bypass Model." *Canadian Journal of Development Studies*, vol. 4, no. 2, 1983: 221-240.

Jolly, Richard, and Colclough, Christopher. "African Manpower Plans: An Evaluation." *International Labour Review*, vol. 106, no. 2-3, August/September 1971: 207-223.

Jolly, Richard. "Future for UN Aid and TA?" *Development*, 1989.

Kirpich, Phillip Z. "Foreign Experts—Their Advantages and Limitations." *Finance and Development*, vol. 4, no. 1, 1967: 44-50.

Kjellstrom, Sven B. "Réforme Institutionnelle et Assistance Technique en Afrique au Sud du Sahara." *Le Mois en Afrique*, 1987: 42-57.

Lecomte, Bernard J. *Project Aid: Limitations and Alternatives*. Paris: OECD, 1986.

Lister, S., and Stevens, M. "Aid Coordination and Management." World Bank, 1992.

Lorenzl, Gunter. "Technical Cooperation in Agricultural Market Development; Experiences from Programmes of the Federal Republic of Germany." *Quarterly Journal of International Agriculture*, vol. 24, April/June 1985: 114-28.

Maddison, Angus. *Foreign Skills and Technical Assistance in Economic Development*. OECD Development Centre. Paris, 1965.

Mann, Amy G., ed. *Institution Building: A Reader*. Bloomington: PASITAM, 1975.

Mikulowski, Witold. "L'Assistance Technique Extérieure et l'Administration des Pays en Développement." *Revue Française de l'Administration Publique*, vol. 26, 1989.

_____. "Les effets réels de la coopération administrative: Quel jugement porter sur le terrain?" *Revue Française d'Administration Publique*, no. 50, avril-juin 1989.

Montgomery, J. *Sources of Administrative Reform: Problems of Power, Purpose and Politics*. Bloomington, Indiana, 1967.

Morss, Elliott R. "Institutional Destruction Resulting from Donor and Project Proliferation in SSA Countries." *World Development*, vol. 12, no. 5/6, 1984: 465-470.

Muscat, Robert J. "Evaluating Technical Cooperation: A Review of the Literature." *Development Policy Review*. London: Sage Publications, vol. 4, March 1986: 69-89.

Mutahaba, Gelase. *Reforming Public Administration for Development: Experiences from Eastern Africa*. African Association for Public Administration and Management. Hartford, Conn.: Kumarian Press, 1989.

Nordic UN Project, The. *The UN in Development, Reform Issues in the Economic and Social Fields, A Nordic Perspective*. Stockholm: GOTAB, 1991.

O'Brien, Rita Cruise. "Colonization to Co-operation? French Technical Assistance in Senegal." *Journal of Development Studies*, October 1971: 45-57.

OECD. *Geographical Distribution of Financial Flows to Developing Countries, 1970-1989*. Paris, 1991.

_____. Development Assistance Committee. *Development Cooperation*. Paris, 1990.

Olson, Craig. "Reflections on Project Organization and Implementation." Chapter Eight in *New Directions in Development: A Study of U.S. AID*. Mickelwait, Donald R.; Sweet, Charles F.; and Morss, Elliott R., eds. Boulder: Westview Press, 1979.

Pavart, F. "Sur Quelques Problèmes de la Coopération Technique." *Afrique et Asie*, no. 50, 2ème trimestre, 1969: 3-19.

Rhomari, Mostafa. "La Coopération Administrative Sud-Sud à l'Epreuve de l'Ajustement." *Revue Française d'Administration Publique*, no. 50, avril-juin 1989.

Robinson, Derek. *Civil Service Pay in Africa*. International Labour Office, 1990.

Schein, Edgar. *Organizational Psychology*. Second Edition. New York: Prentice Hall, 1979.

Schubert, Bernd. "Sustainability of the Effects of Agriculture Projects in German Technical Cooperation: Methodology and Selected Findings of a Cross-Sectional Analysis of 24 Agricultural Projects after Handover to Partner Countries." *Quarterly Journal of International Agriculture*, vol. 24, April/June 1985: 212-41.

"The Sense of Twinning." *West Africa*. 15 December 1986: 2059-2062.

Sufrin, Sidney C. *Technical Assistance: Theory and Guidelines*. Syracuse: Syracuse University Press, 1966.

Swedish International Development Authority. *Development Aid in the 1990s*. Stockholm: Lenanders, Kalmar, 1990.

Swerdlow, I., and Lindeman, J. *The Magnitude and Complexity of Technical Assistance*. Syracuse: Syracuse University Press, 1962.

UNDP and World Bank. *African Economic and Financial Data*. World Bank, 1989.

UNESCO. *Statistical Yearbook*. New York, 1960-1989.

Wallace, Laura. "Reshaping Technical Assistance." *Finance and Development,* 1990: 28.

Winslow, Anne. "The Technical Assistance Expert." *International Development Review*, 4, no. 3, September 1962: 17-24.

World Bank. *The African Capacity Building Initiative Toward Improved Policy Analysis and Development Management*. World Bank, January 1991.

_____. *Social Indicators of Development 1990*. Baltimore: The Johns Hopkins University Press, 1990.

_____. *World Development Report 1990, Poverty*. Oxford: Oxford University Press, June 1990.

_____. Operations Evaluations Department. *Evaluations Results for 1988: Issues in World Bank Lending over Two Decades*. World Bank, April 1990.

_____. *Sub-Saharan Africa: From Crisis to Sustainable Growth. A long-term perspective study*. World Bank, November 1989.

Reports and Papers

General

Algemeen Bestuur Ontwikkelingssamen Wering. "Aktiviteiten Verslag, 1987-1988-1989." Brussels, 1990.

Adamolekun, Ladipo. "Issues in Development Management in Sub-Saharan Africa." EDI policy seminar report no. 19. World Bank, October 1989.

Antoine, Monique Pierre. "Rapports et interactions de l'Assistance Technique et de l'Administration publique." Unpublished note, UNDP, n.d.

Auerhan, J. et al., "Institutional Development in Education and Training in Sub-Saharan African Countries." World Bank Discussion Paper, AFTED, 1985.

Barclay Jr., Albert H. "Successful Adaptation Means Adopting Twins." *Developing Alternatives*. Development Alternatives, Inc., Bethesda, Md., July 1991.

Berg, Elliot. "The Reform of Public Investment Programming in Senegal: An Evaluation." Report prepared for the Projet d'Appui à la Gestion de Développement, Government of Senegal. Development Alternatives, Inc., Bethesda, Md., October 1991.

Birindelli, Marie-H. "Le Cas de la Coopération Technique ACP-CEE: Expériences et Orientations." Contribution au colloque organisé par le Centre Européen de Gestion des Politiques de Développement et l'Africa Leadership Forum." Maastricht, 18-20 octobre, 1991.

Bossuyt, J.; Laporte, G.; and F. van Hoek. "New Avenues for Technical Cooperation in Africa." European Center for Development Policy Management. Occasional Paper. Maastricht, 1992.

Brinkerhoff, Derek W. "Institutional Analysis and Institutional Development: A Survey of World Bank Project Experience." CECPS, World Bank, October 1989.

Bryant, Coralie. "Development Management and Institutional Development: Implications of Their Relationship." American University and Overseas Development Council, October 1985.

Buyck, Beatrice. "The Bank's Use of Technical Assistance for Institutional Development." Working Paper Series no. 578. World Bank, January 1991.

_____. "Technical Assistance as a Delivery Mechanism for Institutional Development: A Review of Issues and Lessons of Bank Experience." Paper prepared for the World Bank's Conference on Institutional Development, December 1989.

Cassen, Robert, and Associates. *Does Aid Work?* Oxford: Clarendon Press, 1986.

Cernea, Michael M. "Nongovernmental Organizations and Local Development." World Bank Discussion Paper no. 40, October 1988.

Cohen, John M. "Expatriate Advisors in the Government of Kenya: Why They Are There and What Can Be Done About It." Harvard Institute for International Development Discussion Paper no. 376, June 1991.

Cooper, F.N. "Comment Améliorer l'Efficacité de l'Assistance Technique en Intégrant le Développement des Capacités dans la Conception des Projets." Seminaire sur la gestion de l'assistance technique, Berlin. Institute du Développement Economique. Mai 1988.

Cooper, Lauren. "The Twinning of Institutions: Its Use as a Technical Assistance Delivery System." World Bank Technical Paper no. 23. World Bank, July 1984.

Davis, T. "Assessment of Institutional Capacity to Effectively Use Technical Cooperation in the Agriculture, Forest, Fishery and National Parks and Wildlife Sectors." UNDP, July 1990.

de Merode, Louis. "Civil Service Pay and Employment Reform in Africa: Selected Implementation Experiences." World Bank, 1991.

Diallo, Aliou; Kelly, Jim; Nikoi, Amon; and North, W. Haven. "Capacity Building for Aid Coordination in the LDCs." Volume 1 of the evaluation report: "UNDP's Role in Capacity Building for the Management of Development Resources in the LDCs." UNDP, 1991.

Eaves, Page. "OED Analysis of Institutional Development." Paper presented to Conference on Institutional Development and the World Bank. World Bank, December 1989.

Esman, Milton. "Strategies and Strategic Choices for Institutional Development." Paper presented at the Conference on Institutional Development and the World Bank, December 1989.

Forss, Kim; Carlsen, J.; Froyland, E.; Sitari, T.; and Vilby, K. "Evaluation of the Effectiveness of Technical Assistance Personnel Financed by the Nordic Countries." DANIDA, Copenhagen, 1990.

Fox, Louise. "Issues in Federal Compensation and Employment" (Brazil). LA1CO. Paper presented at the World Bank Civil Service Reform Seminar, March 1989.

Galal, Ahmed. "Public Enterprise Reform: Lessons From the Past and Issues for the Future." World Bank Discussion Paper no. 119, March 1991.

Gelb, Alan; Knight, John; and Sabot, Richard. "Lewis Through A Looking Glass: Public Sector Employment, Rent-Seeking, and Economic Growth." World Bank, WPS 133, November 1988.

Glentworth, G. "Technical Cooperation in the UK Aid Programme." Paper given at Joint EDI/KIA Seminar on the Management of Technical Assistance, Nairobi, April 17-27, 1990.

Gray, Cheryl. "Institutional Development Work in the World Bank: A Review of 84 Bank Projects." Paper presented at the Conference on Institutional Development and the World Bank, December 1989.

Gregory, Peter. "Somali Civil Service Reform." Paper presented at the World Bank Civil Service Reform Seminar, March 1989.

_____. "Structural Adjustment and Employment in Ghana." Paper presented at the World Bank Civil Service Reform Seminar. August 1988.

Hagan, Kojo, and Silverman, Jerry. "Developing African Consulting Capacity: Proposed Regional Program Strategy Discussion Paper." World Bank, August 1990.

Harari, Denyse. "The Role of the TA Expert: an Enquiry into the Expert's Identity, Motivations and Attitudes." OECD Development Centre Report, 1974.

Heller, Peter S., and Tait, Alan A. "Government Employment and Pay: Some International Comparisons." International Monetary Fund Occasional Paper no. 24. International Monetary Fund, October 1983.

Heller, Peter S., and Diamond, Jack. "International Comparisons of Government Expenditure Revisited: The Developing Countries." International Monetary Fund Occasional Paper no. 69. International Monetary Fund, April 1990.

Hollister, Robinson G. "A Perspective on the Role of Manpower Analysis and Planning in Developing Countries." World Bank Staff Working Paper no. 624, 1987.

Kjellstrom, Sven B., and d'Almeida, Ayite-Fily. "Institutional Development and Technical Assistance in Macroeconomic Policy Formulation: A Case Study of Togo." World Bank Discussion Paper, June 1987.

Klemp, L. "Critical Review of Technical Cooperation." In report of the International Round Table on "The Future of Technical Cooperation." DSE, Berlin, 1989.

Lethem, Francis, and Cooper, Lauren. "Managing Project-Related Technical Assistance: The Lessons of Success." World Bank Staff Working paper no. 586, 1983.

Lindauer, David L. "Government Pay and Employment Policies and Government Performance in Developing Economies." World Bank, WPS 42, August 1988.

Maddison, Angus. "Foreign Skills and Technical Assistance in Economic Development." OECD, Paris, 1965.

Mickelwait, Donald R., and Honadle, George H. "Rethinking TA: The Case for a Management Team Strategy." Development Alternatives Inc., Bethesda, Md., 1984.

Nordic UN Project, The. "Perspectives on Multilateral Assistance, A review based on Eight Case Studies Commissioned by the Nordic UN." Report no. 10, 1989.

Nunberg, Barbara. "Public Sector Pay and Employment Reform." World Bank Discussion Paper no. 68, December 1989.

Nunberg, Barbara, and Nellis, John, with de Merode, Louis. "Civil Service Reform and the World Bank." CECPS. Paper presented at Conference on Institutional Development and the World Bank. World Bank, December 1989.

Oden, Bertil; Egero, Bertil; and Lesteberg, Halvard. "Assessment of Institutional Cooperation as a Form of Aid Based on Statistics Sweden's Cooperation with Central Statistics Office, Zimbabwe and Bureau of Statistics, Tanzania." Swedish International Development Agency, June 1986.

OECD. Development Assistance Committee. "Principles for New Orientations in Technical Cooperation." Paris, 1991.

_____. "Programme Assistance—Use of Programme Assistance By DAC Members: The Quantitative Framework." With a note by the Secretariat. May 1990.

_____. "Program Assistance: Proposed DAC Work and Issues for Discussion at the DAC Meeting at the Level of Senior AID Policy Officials." May 1990.

_____. "Development Assistance: Country Notes for Discussion: Germany." With a note by the Secretariat. May 1990.

_____. "Working Party on Statistical Problems: Technical Cooperation Reporting." With a note by the secretariat. September 1989.

_____. "Sustainability in Development Programmes: A Compendium of Evaluation Experience." September 1988.

Paredes, Ricardo. "Public Sector Employment and Reform in Chile." University of Chile. Paper presented at the World Bank Civil Service Reform Seminar, 1989.

Paul, Samuel. "Institutional Reforms in Sector Adjustment Operations." World Bank Discussion Paper no. 92, June 1990.

_____. "Institutional Development at the Sectoral Level: A Cross Sectoral Review of World Bank Projects." CECPS. Paper presented at the Conference on Institutional Development and the World Bank. December 1989.

_____. "Institutional Analysis in World Bank Projects." Paper for World Bank Conference on Institutional Development and the World Bank. December 1989.

Paul, Samuel; Steedman, David; and Sutton, Francis X. "Building Capability for Policy Analysis." World Bank, WPS 220, 1989.

Perrault, Michel. Hickling Corporation. "Technical Assistance in the Year 2000." Paper submitted to the Canadian International Development Agency, Social and Human Resources Development Division TCD, 1990.

Rajana, M. "Report on Remuneration and Conditions of Service Offered by Expatriate Agencies Operating in Mozambique." UNDP, March 1991.

Roberts, Lee. "The Policy Environment of Management Development Institutions in Anglophone Africa, Problems and Prospects for Reform." EDI policy seminar report no. 26. World Bank, 1990.

Robinson, Trevor. "Ghana Civil Service Reform Programme." In U.K./ODA, *Civil Service Reform in Sub-Saharan Africa*. Record of a Conference organized by ODA, March 1989.

Sandberg, Bengt G. "Report on Technical Assistance in SSA." Storrar report, 6th edited and condensed version. World Bank, 1982.

Sao Tomé and Principe. Round Table Conference document, vol. II, ch. IV. "Technical Assistance: Situation and Perspectives." 1992.

Sayed, K.S., and Baum, Warren. "The Consulting Profession in Developing Countries; A Strategy for Development." World Bank, WPS 733, July 1991.

Shirley, Mary, and Nellis, John. "Public Sector Management Reform, Lessons of Experience." World Bank, 1990.

Silverman, Jerry M. "Public Sector Decentralization: Economic Policy Reform and Sector Investment Programs." World Bank African Region Public Sector Management Division. Division study paper no. 1, November 1990.

_____. "Technical Assistance and Aid Agency Staff: Alternative Techniques for Greater Effectiveness." World Bank Technical Paper no. 28, 1984.

Steedman, David W. "The Institutional Dimensions of Investment Lending: Education and Power Projects in FY 1987-88." Paper presented at the Conference on Institutional Development and the World Bank, 1989.

_____. "Capacity Building for Policy Analysis in Sub-Saharan Africa: A Review of the Experience of Selected Donors." ARA Consultants, April 1987.

Stevens, Michael. "Bank Experience and Emerging Issues in Public Expenditure Management." CECPS. Paper presented at the Conference on Institutional Development and the World Bank, December 1989.

Sullivan, Roger. "Institutional Development Work in the World Bank: Management Systems, Organization and Incentives." AFTPS. Paper presented at the Conference on Institutional Development and the World Bank, December 1989.

Sunshine, R.B. "La Situation Actuelle dans le Domaine de l'Assistance Technique: Une Synthèse de l'Expérience Récente de la Banque." World Bank, April 1985.

Swedish Ministry for Foreign Affairs. "Swedish Development Cooperation with Sub-Saharan Africa in the 1990s." Stockholm, January 1989.

UNDP. "Technical Cooperation in the Development of the Least Developed Countries." Report for the second United Nations Conference on the Least Developed Countries. 1989.

_____. "Programme and Projects Manual, Overall Framework of UNDP Mandate." February 1988.

_____. "Implementation of New Dimensions in Technical Cooperation: Report by the Administrator." October 1976. Prepared for the twenty-third session of the Governing Council.

_____. "15 Years and 150,000 Skills: An Anniversary Review of the United Nations Expanded Programme of Technical Assistance." Technical Assistance Board, 1965.

United Kingdom, Overseas Development Administration. "British Aid Statistics, 1980-1984 and 1985-1988," Government Statistical Service, 1985 and 1990.

_____. "Civil Service Reform in Sub-Saharan Africa." Record of a Conference Organized by the Overseas Development Administration, March 29-31, 1989.

_____. "Changing Patterns in UK Technical Cooperation." Note, delegation for the DAC Meeting on Technical Cooperation, November 15-16, 1988.

United Nations. "Report of the High Level Committee on the Review of Technical Cooperation Among Developing Countries." 1987.

U.S. Agency for International Development. "AID Policy Paper: Institutional Development." Bureau for Program and Policy Coordination, 1983.

Vrancken, Fernand. "Technical Assistance in Public Administration: Lessons of Experience and Possible Improvements." International Institute of Administrative Sciences, Brussels. XIIth International Congress of Administrative Sciences, Vienna, July 16-20, 1962.

Wescott, Clay; Haregot, S.; and Coker, I. "Rationalization of Financial Support to the Civil Service of Uganda. UNDP, May 1990.

Wheeler, Joseph. "Technical Assistance From the Point of View of the Donors." Paper prepared for the 9th Meeting of the U.N. Inter-Agency Task Force on Africa (UNPAAERD). Paris, May 1989.

World Bank; United Nations Development Programme; and Development Assistance Committee. Inter-agency Task Force: "Technical Cooperation in Africa: Issues and Concerns for Improving Effectiveness." 1989.

World Bank. "Managing Technical Assistance in the 1990's: Report of the Technical Assistance Review Task Force." 1991

_____. "Terms of Reference Guidelines for Technical Assistance." March 1991.

_____, Operations Evaluation Department. "Evaluation Results for 1990." 1991.

_____. Africa Region. "A Framework for Capacity Building in Policy Analysis and Economic Management in Sub-Saharan Africa." March 1990.

_____. "Annual Technical Assistance Report." Paper prepared by the Central Operations Department, 1990.

_____. "Bank-Financed Technical Assistance Activities in Sub-Saharan Africa." December 1989.

_____. "Technical Assistance as a Delivery Mechanism for Institutional Development: A Review of Issues and Lessons of Bank Experience." December 1989.

_____. Operations Evaluation Department. "Free-Standing Technical Assistance for Institutional Development in Sub-Saharan Africa." World Bank report no. 8573, April 1990. (Country studies also available.)

_____. Africa Region. "Building Capacities in Economic Management and Policy Analysis in Sub-Saharan Africa." 1988.

_____. "Improving Design and Delivery of Technical Assistance in Africa." March 1987.

_____. Central Operations Department. "World Bank Technical Assistance Activities and Issues: Fiscal years 1982-1986." 1987.

_____. "Report on the West Africa Region Task Force on Technical Assistance and Training Effectiveness." November 1986.

_____. "Improving the Effectiveness of Aid Coordination in Sub-Saharan Africa." Paper prepared for the Development Committee. 1986.

_____. Projects Policy Department. "Issues Paper. Team Approach: Using External and Local Consultants." November 1985.

_____. "Report on Technical Assistance in Sub-Saharan Africa." August 1982.

_____. Operations Evaluation Department. "Review of Training in Bank-Financed Projects." Report no. 3834, March 1982.

NaTCAP Papers

Bénin, République Populaire du. "Document 'O'. De Politique de Coopération Technique," May 1989.

Burkina Faso. "Coopération Technique, Document d'Orientation," 1991.

_____. NaTCAP. "Présentation Succincte à l'attention de M. le Ministre du M.P.C.," Novembre 1990.

_____. NaTCAP diagnostic study. "Contraintes à la Relève," Septembre 1990.

Burundi, Gouvernement de la République du. "Cadre d'Orientation Politique de la Coopération Technique," Février 1990.

_____. "Document d'Orientation sur la Coopération Technique," 1990.

_____. "L'Assistance Technique: Cadre d'Orientation Politique," 1989.

_____. "Rapport du Ière Séminaire National sur l'Assistance Technique," Septembre 1988.

Gambia, Government of. "NaTCAP, Gambia TCPFP." UNDP, June 1990.

Geli, Paul. "Review of Technical Cooperation Policy Framework Papers." Regional Bureau for Africa, UNDP, October 1991.

Ghana. "NaTCAP, Ghana TCPFP." Ministry of Finance and Economic Planning. UNDP, October 1990.

_____. "Technical Cooperation Policy Framework Paper," 1989.

Gouvernement Haitien/PNUD. "Quelle assistance technique pour Haiti?" Juin 1990.

Government of Guinea-Bissau. "Situation et perspectives de l'assistance technique en Guinée Bissau." Document de travail présenté à la Conférence de Table Ronde, Genève, juin 1988.

Gudgeon, Peter. "Technical Cooperation and the Public Investment Programme," 1990.

Guinée, République de. "Programme de Coopération Technique: 1990-1992," Avril 1990.

_____. "La Politique de Coopération Technique," Avril 1990.

Guinée Bissau, Republica da. "Documento do Governo Sobre a Politica De Cooperaçao Téchnica," March 1990.

Honadle, George H. "Harnessing Technical Cooperation for Development: A Review of the NaTCAP Effort of the UNDP," vols. I and II, June 1990.

Kapur, Shiv S. "NaTCAP Methodology: An Issues Paper." Paper prepared for the Regional Bureau for Africa, 1987.

Malawi, Government of. "Statement of Policy on Technical Cooperation for Human and Institutional Development in Malawi," May 1990.

Mali, République du. "Document d'Orientation sur la Politique de Coopération Technique du Mali," 1991.

Mozambique, Republica Popular de. "Document de Orientacao Sobre Politica de Cooperacao Tenica." Versao de trabalho, October 1989.

Niger, République du. "Document d'Orientation de la Coopération Technique (DOCT)," March 1991.

Olson, Craig, and Chadouet, Jean. "NaTCAP Evaluation Mission: Chad," 1990.

Omolo-Opere, Valentine. "UNDP Technical Cooperation: Policies and Practices." Paper for delivery at an EDI/UNDP/KIA seminar in Nairobi on Management of Technical Cooperation, April 1990.

République Centrafricaine, Ministère du Plan, des Statistiques et de la Coopération Internationale. "Evaluation et Programmation Nationale de la Coopération Technique." Deuxième version. UNDP, juillet 1986.

Soos, Helen, and Mbeya, Finias. "Optimizing the Utilization of Technical Cooperation: An Assessment of Institutional Capacity in the Industrial Sector of Malawi." Prepared for NaTCAP, UNDP, November 1990.

Tanzania. "TCPFP for Tanzania." NaTCAP, UNDP, June 1991.

Tchad, République du. "Document-Cadre d'Orientation de la Coopération Technique," Juin 1990.

––––––. "Etude sur les Procédures de Programmation et de Gestion des Projets de Coopération Technique," Avril 1990.

––––––. "La Coopération Technique dans le Système de Formation des Cadres," Février 1990.

———. "Etude Diagnostique de la Coopération Technique au Tchad," Février 1990.

Togo. "Grandes Orientations du Togo en vue de la Formulation d'une Politique Nationale en Matière de Ressources Humaines dans la Coopération Technique," Mai 1991.

UNDP and World Bank. "Annual Report on the Technical Cooperation Management Project: Swaziland," 1990.

———. "Somalia: Report of a Joint Technical Cooperation Assessment Mission," 1985.

UNDP. "Pour un Renforcement des Capacités de Gestion et une Meilleure Programmation de la Coopération Technique au Gabon." Version finale du rapport de la Mission préliminaire NaTCAP/MDP, mars 1991.

———. "Report of the NaTCAP Evaluation Mission to Guinea," 1991

———. "Guinée, Assistance à la Réforme Administrative." Rapport de la mission d'évaluation, octobre 1990.

———. Regional Bureau for Latin America and the Caribbean. "The Programme Modality in UNDP: A Case Study and Notes on its Application," 1990.

———. "Discussion Paper on Technical Cooperation." Prepared for Uganda Consultative Group Meeting. UNDP, 1990.

———. "A Review of UNDP's Development Cooperation with the LDCs and Proposals for the 1990s," 1990.

———. "Seminar on National Technical Co-operation Assessment and Programmes (NaTCAP)." Report of the Seminar, Regional Bureau for Africa, September 1989.

———. "Technical Cooperation in African Development: An Assessment of its Effectiveness in Support of the UNPAAERD 1986-90," 1989.

———. "Technical Cooperation in Ghana; Issues and Policies." Paper presented by the UNDP, March 1989.

———. "NaTCAP Malawi—Master Plan," 1989.

_____. "Plan Directeur." Exercice NaTCAP/Cap Vert, May 1989.

_____. "NaTCAP Data Requirements and Standard Tables," September 1988.

_____. "Improving Technical Cooperation in Uganda," Consultative Group Meeting for Uganda. 1988.

_____. "Strengthening Technical Cooperation in Tanzania" Consultative group meeting for Tanzania, July 1988.

_____. NaTCAP. Central African Republic, 1988.

_____. "Technical Assistance Policies and Requirements in Mozambique," 1988.

_____. "Round Table Conference for Lesotho: Technical Cooperation Paper," October 1988.

_____. "Swaziland Technical Cooperation and Development Prospects and Challenges." Executive summary, a technical cooperation strategy for Swaziland. New York, 1986.

Williams, Maurice J. "Evaluation of NaTCAP." Regional Bureau for Africa, February 1991.

Williams, M., and Nikol, A. "Final Report, NaTCAP." UNDP Evaluation Mission, Malawi, December 12, 1990.

Background Papers Done by and for the Regional Bureau for Africa, UNDP

Adei, Stephen. "Overview of Technical Assistance/Cooperation," 1990.

de Bernis, Marc. "La Coopération Technique: Une Variable Macro-Economique," n.d.

_____. "Une Approche Macro-Economique de la Coopération Technique en Afrique." UNDP, September 1990.

_____. "La Coopération Technique: de Quoi Parle-t-on?" UNDP, September 1989.

_____. "L'Assistance Technique en Afrique: Données Quantitatives," 1988.

Fukuda-Parr, Sakiko. "The Expert-Counterpart Model of Technical Cooperation: Debunking the Myths," 1990.

_____. "Current Policy Issues in Technical Cooperation." Seminar for UNDP field economists, 1990.

Havnevik, Kjell J. "Unequal Partners: The Role of Donors and Recipients in the Identification, Design and Implementation of Technical Cooperation Projects," May 1990.

Jenks, Bruce. "Towards a Programme Approach," January 1990.

Laranjeiro, Maria Jose. "Intégration des Projets de Coopération Technique dans le Processus Budgétaire et de Planification Nationale," 1990.

Lopes, Carlos. "Assessment and Programming of Technical Cooperation: NaTCAP Experience," 1990.

_____. "Une Perspective Historique de la Coopération Technique en Afrique," 1990.

Sandberg, Bengt. "A Programme Approach." Regional Bureau for Africa, UNDP, 1991.

_____. "The Role of Host Country Environment in Successful Technical Cooperation," 1990.

_____. "Notes on the Budgeting of Technical Assistance," 1990.

INDEX

Administrative environment
    Donor governments efforts to improve, 209; by employment reduction, 214-15; by general administrative reforms, 219-20, 232; by pay reforms, 215, 217-19; by salary supplements to local employees, 209-14
    For effective capacity building, 196, 203; civil service reforms for, 230-34; downsizing public sector for, 235-40; factors influencing, 197; salary supplement reforms for, 221-30
    Management deficiencies: from civil service problems, 199-202; from inadequate wage incentives, 200, 203-04, 207, 209
    Proposed reforms for more effective, 268-69
    Recipient governments efforts to improve, 209-10
    See also Civil service; Public sector; Working environment

African Capacity Building Foundation, creation, 58

Agency for International Development (A.I.D.), agricultural research projects, 2, 19

Agriculture
    Productivity: factors influencing slow growth, 1; technical cooperation and, 1-2
    Research: development impact of, 17; technical cooperation and, 19-20
    TA personnel in, 34(14n)

A.I.D.. See Agency for International Development

Balance of payments support. See Budget support

Belgium, financing of technical cooperation, 79, 81

Benin, TCPFP on effectiveness of technical cooperation in, 5

Botswana, gap-filling personnel used in, 111-12
Budget support
    Development assistance in form of, 48
    Function, 49
    Objective, 49
Burkina Faso
    Expatriate TA personnel, 7; in agriculture and hydrology, 34(14n)
    Technical cooperation to: effectiveness, 5, 6; ultimate objective, 59
Burundi
    TA personnel, 7; expatriate medical, 77, 263
    Technical cooperation to, 81; comprehensive programmes for, 154, 177

Cameroon, capacity-building projects in education, 25
Capacity building
    Colonial rule and, 69-70
    Defined, 60, 63, 89(16n)
    Donor versus recipient management effect on, 139
    Institutional building compared with, 63
    Main activities of, 62-63
    Projects in education, 28
    Proposed administrative reform strategies for effective: civil service reforms, 230-34; downsizing public sector through privatization, 235-40; further consideration of salary supplement issues, 221-30
    Technical cooperation for: Cassen Report on, 18, 21; criticism of, 15; focus on 58-60, 63; requirements for, 196; World Bank on, 24
    Work environment impact on, 196-98, 203
Capital assistance
    Defined, 50, 52
    Distinction between technical cooperation and, 48-49, 50, 52
    Financing of infrastructure, 50, 52
    World Bank and, 24
Cassen Report, evaluation of technical cooperation, 18-21, 28
Central African Republic
    Civil service wages, 209
    National TA personnel, 6
    Technical cooperation to, 81

Chad
  Importance of technical cooperation for training and knowledge transfer, 59
  Technical assistance personnel, 8; contribution to project success, 6; ratio of expatriate to national, 76-77
Civil service
  Difficulties in reforming, 230-31
  Donor country employment and pay reforms for, 210-15, 217-18
  Piecemeal versus enclave approach to, 232-34
  Shortcomings: ministerial turnover, 202; number, 200; personnel management, 201; responsibilities versus resources, 201-02; training system, 202; unrealistic budgets, 200-01; wage inadequacies, 200, 203-04, 207, 209
Colonial rule
  Negative aftereffects on capacity building, 69-70
  Restraints on education and skill development, 64
  Technical cooperation effects during, 65
Congo, technical cooperation to, 81
Costs, technical cooperation, 14
  Criticism of, 245
  Divergence between user and payer, 170-71
  Donor versus recipient country, 168
  Management process consideration of, 177
  Nuisance, 170, 192(4n), 193(8n)
  Opportunity, 165; financial costs versus, 167; in trade-off with other development assistance, 168-69; transaction, 168, 192(4n)
  Role in market supply and demand determination, 165, 171-74
Côte d'Ivoire, technical cooperation to, 80-81, 178

DAC. See Development Assistance Committee
Decision-making process, 128
  Capital investment projects, 129-31
  TC projects, 131-33
Delivery system, technical cooperation project
  Failures in project identification and design, 30; in defining project objective, 97; lack of local participation, 94-95; reliance on expatriate personnel, 95-96; in time for project preparation, 96-97; in use of project mode, 97-98
  Implementation weaknesses, 90, 98; from deficiencies in administrative and work environments, 98-99; from personnel recruitment and supervision problems, 99-100
  Institutional twinning relationship to, 116-20

Proposed reforms for, 121; improved personnel recruitment and training, 107; increase in donor time for project design and planning, 106-07; weaknesses in, 108-09

Development assistance
- Forms, 48
- Objectives, 49-50
- Percent of gross national product, 127
- Problems in classifying activities under, 50-51
- To Sub-Saharan Africa, 68, 127
- Technical cooperation share of, 68, 72, 127

Development Assistance Committee (DAC)
- Distinction between technical cooperation and technical assistance, 47, 53
- Evaluation of technical cooperation, 36(28n), 244; in education, 20, in health, 20-21
- Funding of African higher education opportunities, 67
- On importance of capacity building, 58
- Proposed changes in technical cooperation, 244; on comprehensive programming approach, 151; on donor salaries to civil service, 225-26, 227; for market improvements, 166; for TC delivery system, 107, 113, 115, 123-24(13n)
- Questionnaire on salary supplements, 213
- Source of data on technical cooperation, 71

Djibouti, technical cooperation to, 81

Education
- Capacity-building projects in, 25
- Colonial rule restraints on, 64, 65-66
- Expansion, <u>1950s</u>, 65
- Foreign universities for higher, 67
- Independent African states high priority on, 66-67
- Skill development problems, 60
- Technical cooperation role in, 20, 67-68

Ethiopia, technical cooperation to, 81

Evaluation of technical cooperation
- Cassen Report, 18-21
- Conclusions, 28-32
- Forss Report, 21-24
- International conferences for, 35(23n)
- Methodology problems in, 16-18
- UNDP/World Bank, 2-3
- <u>See also</u> National Technical Cooperation and Assistance Programmes; Technical Cooperation Policy Framework Papers

Expenditures, TC projects, 50

Government revenues and export earnings compared with, 84
Input composition: for equipment and operating costs, 84; for training programs, 84-85
OECD understatement of, 71, 83
Rules for, 51

FAO. See Food and Agriculture Organization
Financing of technical cooperation
    Bilateral versus multilateral, 68, 79-81
    Categories, 71
    For personnel, 76, 86
    Share of all development assistance, 68, 72
Food and Agriculture Organization (FAO), 2
    On agricultural research institutions, 19
Foreign exchange, technical cooperation as source of, 83
Forestry, technical cooperation and, 19
Forss Report, evaluation of technical cooperation, 21-24
France
    Colonial rule: inhibition of technical skills and management competence, 64; scholarships for foreign students, 65
    Financing of technical cooperation, 79, 80-81
    TA personnel provided by, 51, 68; *assistance technique* provision for, 43, 87(1n); decline in number of, 76; described, 91(35n); phasing out of substitution-type, 54
Free-standing technical cooperation projects
    Described, 131
    For institutional development, 52
    Management requirements for, 132
    Number of, 132
    Potential management reforms for, 159
    "Soft," 24, 52-53

Gabon
    Payments to TA personnel for housing, 193(6n)
    Technical cooperation to, 81
Gambia
    Civil service wages, 207, 218
    Downsizing of public sector by privatization, 235
    TCPFP on skill transfer and institutional building, 5
Germany, financing of technical cooperation by, 79, 80, 81

Ghana
　　Availability of local consulting personnel, 115
　　Civil service: enclave approach to, 233, 234; reduction in employment, 215; wages, 207
　　Institutional twinning, 118
　　Skills Mobilization Scheme, 229-30, 234
　　TCPCP: definition of technical cooperation 59; on planning TC projects, 8
Grants, technical cooperation
　　Bilateral, 79-82
　　Costs, 168, 245
　　From former colonial powers, 80-81
　　Multilateral, 79, 81
　　Recipient countries preference for, 193(9n)
　　Trends in, <u>1970-89</u>, 71-72
Great Britain. <u>See</u> United Kingdom
Gross national product, development assistance share of, 127
Guatemala, institutional twinning project, 119
Guinea
　　Comprehensive programming for technical cooperation, 154, 156, 177; effect on TA personnel, 178
　　Reduction of civil service employment, 215
Guinea-Bissau
　　Comprehensive programming for technical cooperation, 154, 177
　　Number of expatriate TA personnel, 76, 102

Haiti, technical cooperation in, 122(1n)
Health projects, technical cooperation and, 20-21

ILO. <u>See</u> International Labour Organization
IMF. <u>See</u> International Monetary Fund
Independence, African states
　　Status of education upon, 65-66
　　TA personnel serving after, 68
Institutional development
　　Capacity building compared with, 63
　　Defined, 60, 89(13n)
　　Elements, 61-62
　　Extent of Sub-Saharan Africa, 14
　　Negative aftereffects of colonial rule on, 69-70
　　Organizational development compared with, 61
　　Technical cooperation for: Cassen Report on, 18-21; challenges to, 29-30; emphasis on, 58-60; evaluation of, 12, 15, 25-26; by

improving administration, 50; operational support TC compared with, 53-54, 55, 58; personnel for, 55; World Bank Report on, 24-28

Institutions
    Defined, 60
    Organizations compared with, 60-61
    Role in economic growth, 59
    See also Institutional development; Twinning, institutional

International Labour Organization (ILO), capacity-building project in education, 25

International Monetary Fund (IMF), 174, 175
    Ceilings on wages, 228
    Civil service reform programmes, 217, 218

Investment-related technical cooperation 71
    Aid activities classified as, 51
    For capital projects, 47, 48-49; donor initiated, 130; programming processes for, 130-31; screening procedures, 130
    Percent of total technical cooperation, 71
    Personnel for, 47
    Public, 130, 177-78

Italy
    Agricultural research assistance to Somalia, 2
    Financing of technical cooperation, 80

Jamaica, structural adjustment loan to, 218
Japan, financing of technical cooperation, 80, 81

Kenya
    TA personnel, 95; case study of, 22, 23; local consulting by, 115
    Technical cooperation to, 81
Knowledge, technical assistance for transfer of, 50, 196

Labor market, Africa
    Availability of trained personnel, 6-7, 9-10, 77
    For effective capacity building, 196
    Salary supplements effect on internal, 223-24
    Strategy for success in, prior to <u>1980,</u> 196-97

Lesotho
    Free-standing TC projects in, 132
    Study of nationals to replace expatriate personnel, 35(20n)

Loan-financed technical cooperation
    Grant-financing to replace, 193(9n)
    Institutional strengthening as condition for, 174

World Bank, 71, 134; assessment of, 26-28; for hard projects, 24; on reimbursable terms, 168; sectoral and structural adjustment loans, 27, 169, 218, 227

Maastricht Conference, on technical cooperation, 13
    Case study findings on management, 161(10n)
    On NaTCAP approach to management, 163(22n)
    Reform proposals for management, 140-41
Madagascar
    Capacity-building projects in education, 25
    Local consulting personnel, 115
    Technical cooperation to, 81, 82
Malawi
    Capacity for TC management, 163(24n)
    Expatriate versus national personnel, 14
    Free-standing TC projects, 132
    Human resources inventory in, 163(30n)
    Policy document on technical cooperation, 59
    Vacancy rates for professional and technical grade jobs, 102
Mali
    Capacity-building projects in education, 25
    National TA personnel, 6
Management, technical cooperation, 8, 9
    Comprehensive programming approach for, 151-52, 251-52; advantages, 153, 155-56; countries using, 154; disadvantages, 154-55, 258
    Defined, 39(55n)
    Diversity in project requirements for, 132-33
    Methods for increasing national role in: evaluation of, 257-58; NaTCAP plans for, 148-51, 251; by programming approach, 251-52; by voluntary donor authority transfer, 147-48, 250
    NaTCAP approach to stronger, 148; achievements of, 150-51; development of management information system, 149; focus on institutional requirements, 150; TCPFP for, 149-50
    Problems of donor dominance in, 31, 127-28; from administrative deficiencies in recipient countries, 136-38; from donor choice of personnel, 138; from implementation control, 134; on project design and ideas, 133-34; proposed donor coordination to handle, 139-40; from recipients refusal to take on responsibilities, 162(13n)
    Project-by-project approach to, 151; deficiencies in, 152-53; proposed replacement by programme approach, 251
    Proposals for national, 141-42, 145-47

Reform strategies: changes in environment surrounding technical cooperation, 160; gradualist, 156, 158; manpower assessments, 158; at micro level, 158-59; priorities for handling problems, 157-58

Market, technical cooperation
    Drawbacks to TC programming approach, 188; concern over donor funding schedule, 190; monitoring of disbursements, 190; project programming and costing, 189; uncertainties and risks, 191
    Elements; build on TC programmes, 175, 177-79; introduce prices, 179-81; reduce restrictions on hiring of nationals, 184-88, 192; separate treatment for supplies and personnel, 182-83; training strategies, 183-84
    Imperfections, 165-66, 264-65; absence of cost consciousness, 167-68, 173-74; divergence between user and payer costs, 170-71, 191-92; price inelasticity on supply side, 171-74; recommended policy reforms for, 265-68; restrictions on buyers, 174-75; uncertain opportunity costs, 168-70, 174;

Meteorology, technical cooperation and, 19

Mozambique, comprehensive programming for technical cooperation, 154, 177

Muscat, Robert, evaluation of technical cooperation, 18-21, 28

National Technical Cooperation and Assistance Programme (NaTCAP), 4
    Approach to stronger TC management, 148-51
    Comprehensive programming approach to technical cooperation, 151, 154, 155, 156, 177
    Improved data systems through, 82
    Surveys: on number of TA personnel, 76, 84, 100; on number of TC projects, 132
    On TC ineffectiveness, 244
    See also Technical Cooperation Policy Framework Papers

Netherlands, financing of technical cooperation, 79, 80, 81

Nigeria, local consulting personnel, 115

Nordic Conference on Technical Cooperation, 95

OECD. See Organization for Economic Co-operation and Development

Operational technical cooperation, 14
    Institutional development versus, 53-54, 55, 58
    Substitution-type gap-filling personnel for, 8, 55, 58; need for, 111; problems with, 112; use, 111-12

Organization for Economic Co-operation and Development (OECD)
    Data on TA personnel, 72
    Data on TC aid, 82
    On resident expert-local counterpart model, 101
    TA personnel in Sub-Saharan Africa from, 68
    See also Development Assistance Committee
Organizations
    Defined, 60-61
    Development; capacity building versus, 63; institutional development versus, 61

Papua New Guinea, gap-filling personnel used in, 111
Personnel, technical assistance
    Capacity building versus operational roles, 29
    Described, 47
    Expatriate: budgetary limits on use of, 7; cost, 76, 100, 245; criticism of long-term, 29; expansion of, 68; financing, 76, 86; proposed donor contributions to fund for, 187-88; recipient country role in recruiting, 8; reliance on, 5-6, 95-96; source of, 77
    Forss Report on effectiveness of, 21-24
    Introduction of market prices for, 180-81
    Knowledge transfer by, 50
    Nationals; availability of trained, 6-7, 9-10, 77; recommended reforms relating to, 263-64
    Number: increase in, upon independence, 68; new technology influence on, 86-87; overstatement of, 72, 76
    Substitution-type, gap-filling, 54-55; output, 195-96; problems with, 112, 113; reform proposal for increasing, 263-64; use, 111-12
    TCPFPs recommendations on, 7-8
    See also Resident-expert local counterpart model
Population, technical cooperation impact on policies for, 21
Portugal, financing of technical cooperation, 81
Principe, payments to TA personnel for housing, 193(6n)
Productivity
    Agricultural, 1-2
    Capital assistance and technical cooperation to increase, 50
Programmes
    Public investment, 130, 177-78

TC comprehensive management, 140; advantages, 150, 155-56; cost awareness under, 177, 178; countries using, 154; disadvantages, 154-55, 258; need for improvements in, 179; recommendations for, 151-52

Project-by-project approach
    Criticism of, 152-53
    Proposed replacement by programme approach, 151, 153

Public sector
    Downturn in operating efficiency of, 199
    Proposed privatization to downsize, 235-38, 270
    See also Administrative environment; Civil service; Wages and salaries

Reform proposals, technical cooperation
    Consensus: on donor delivery package for project, 247, 255; on improving work environment, 252-54, 258-62; on reduced reliance on resident expatriate personnel, 248-49; on strengthening local management, 249-52, 257-58
    Imperfections in, 270
    Nonconsensus: for administration, 268-69; for downsizing public sector, 270; for market, 264-68; for TA personnel, 263-64
    Urgency of, 271

Resident expert-local counterpart model
    Criticism of, 100-01, 247-48; personnel recruitment and loyalty problem, 103-04; as training vehicle, 104-06; weaknesses in meeting basic preconditions, 101-03
    Elements, 100
    Proposed reforms in, 100, 121; by adoption of coaching models, 110, 114-15; by replacing gap-filling personnel, 111-13; by use of local consultants, 115-16
    Proposed substitutes for, 249; problems with, 256-57

Round Tables and Consultative Group
    Donor programming reforms, 139-40
    TCPCP distribution to, 149

Rwanda, technical cooperation to, 81

Salaries. See Wages and salaries
SALS. See Structural adjustment loans
Sao Tomé, payments to TA personnel for housing, 193(6n)
Sectoral loans (SECALs), 27, 227
Senegal
    Civil service salaries, 209
    Expatriate TA personnel in, 77

French TA personnel in, 70
French wages for administrators in, 64
Public investment programmes, 132
Technical cooperation to, 81
Sierra Leone, capacity-building projects in, 25
Skills, TA personnel, 77
Capacity building with upgrading of, 62-63
Donor country efforts to improve, 214, 215
Training programs to transfer, 29
Somalia
External assistance in agricultural research, 2
Salary supplements impact in, 241(23n)
Technical cooperation to, 81
Structural adjustment loans (SALs), 27, 169, 218
Sub-Saharan Africa, 42
Civil service shortcomings in, 199-202
Development assistance to, 68, 127
Institutional development projects, 25
Institutional growth, 14
TA personnel in, 68, 72, 76-77
TC grants to, 71-72
Technical cooperation as public resource, 83-84
Sudan, technical cooperation to, 81
Swaziland, comprehensive programming for technical cooperation, 154, 177, 178
Sweden
Financing of technical cooperation, 80, 81
Use of institutional twinning, 117, 120

TA. See Technical assistance
Tanzania
Technical assistance personnel; case study, 22; as controllers, 95; cost, 14
Technical cooperation to, 81
TC. See Technical cooperation
TCPFPs. See Technical Cooperation Policy Framework Papers
Technical assistance (TA)
Defined, 33(1n), 37(43n); country and organization differences in, 43
Distinction between technical cooperation and, 47
Free-standing, 27, 51, 54; for institutional development, 52; number and size of projects, 131; problems in implementing, 173; soft, 24, 52-53

Hard, 24; engineering and scientific knowledge for, 53; investment-related, 52; output from, 195-96; success, 27
Investment-related, for capital projects, 48-51
Lending operations, 26-28
Operational support versus institutional development, 55
See also Personnel, technical assistance; Technical cooperation

Technical cooperation
Bilateral versus multilateral, 68, 79-81
Defined, 33(1n); country and organizational differences in, 43
Early expectations for, 1-2
Ineffectiveness, 3-4, 11-12; in achieving recipient country self-reliance, 244-45; from costliness, 4, 245; donor assessments of, 10-15; from poor planning and management, 8-10, 30-32, 82; sources of 246-47; variations in, 78
Inputs, 28, 48, 84-85
Institution-building, 12, 14-15, 24-26, evaluation of, 28, 29-30
Objectives, 5, 48, 49; effectiveness measured against, 15-16; knowledge transfer as, 48, 50; self-reliance as, 49, 50
Role in capacity building, 18-21, 24, 63; evaluation of, 28
See also Capacity building; Costs, technical cooperation; Evaluation of technical cooperation; Grants, technical cooperation; Investment-related technical cooperation; Management, technical cooperation, Market, technical cooperation; Reform proposals, technical cooperation

Technical Cooperation Policy Framework Papers (TCPFPs)
Consensus position on TC reform, 149, 151
Described, 4-5
Recommendations: on central government planning for TC projects, 8-10; on use of TA personnel, 7-8

Training, personnel
Fellowships for, 184
Formal versus informal, 29
In-service, 183-84
On-the-job, 29
Programmes for, 84, 85, 202

Twinning, institutional
Advantages, 118-19
Cost, 119-20
Disadvantages, 119, 256-57
Experience with, 117
Purpose, 116-17

Uganda
- Civil service wages, 239(5n)
- Expatriate versus national TA personnel, 6-7
- Proposed common fund for pay supplements, 229
- TCPFP, 9
- Technical cooperation to, 81
- United Kingdom public administration program for, 240(16n)

UNDP. See United Nations Development Programme

Unequal partnership, 160(4n)

UNESCO. See United Nations Economic, Social and Cultural Organization

UNICEF. See United Nations Childrens Fund

United Kingdom
- Colonial rule: university education under, 64; funds for technical cooperation under, 65
- Financing of technical cooperation, 79, 81
- Number of TA personnel, 68, 76
- Overseas Development Administration report on Sub-Saharan civil service, 199
- Public administration program in Uganda, 240(16n)
- Use of institutional twinning, 116, 124(19n)

United Nations
- Expanded Programme for Technical Assistance, 65
- TC resources from, 1960s, 68

United Nations Childrens Fund (UNICEF), on technical cooperation, 13-14

United Nations Development Programme (UNDP)
- Evaluation of TC projects, 20
- Financing of TC projects, 51, 81; recommended reforms for, 254
- On institutional twinning, 116-17
- Policy on donor compensation to government personnel, 225
- On resident expert-local counterpart model, 101
- Sponsorship of NaTCAP approach to TC management, 148, 150
- TC assessment, 2-3, 11-12

United Nations Economic, Social and Cultural Organization (UNESCO)
- Capacity-building projects in education, 25
- Evaluation of technical cooperation in education, 20

United States
- Financing of technical cooperation, 65, 79, 81
- TC resources from, 1960s, 68

Wages and salaries
    Donor country payment of supplementary: arguments for, 212, 222, 225-26; differentials in, 210; disadvantages, 213-14, 223-24, 229; proposed common fund for, 228; proposed reforms for, 221-22; recipient countries reaction to, 239-40(8n); recommended guidelines for, 226-28; restrictions on, 212, 240(9n); by type of employee, 210-11; uncertainty over treatment of, 230; vehicle for delivery of, 211-12
    Donor country reforms for, 214-15; impact, 217-18; problems with, 218-19
    Public sector, 195; decrease in, 204, 207; differentials, 207, 209; growing divergence between effort and, 207; policies governing, 203-04
Working environment, 32
    Absence of congenial, 221
    Consensus reform proposals for, 252-53, 258-62
    For effective capacity building, 196-98, 203
    Proposed strategy for improving: by changing donor government salary supplements, 221-20; by civil service reform, 230-34, 252-53; by downsizing public sector, 235-38
    See also Administrative environment
World Bank
    Assessment of technical assistance in agricultural research, 2-3
    Civil service reform programme, 214, 217, 218
    On comprehensive programme for technical cooperation, 156
    Donor programming reforms, 139-40
    Drawbacks as provider of technical cooperation, 109-10; proposed reforms to correct, 123(9n), 123-24(13n)
    Evaluation of African institutions, 60
    On resident expert-local counterpart model, 101
    Sectoral and structural adjustment loans, 27, 218, 169, 227
    TA review Task Force, 26-28, 166
    Technical assistance: amount of, 79; for capacity building, 24, 25; evaluation of projects, 25-28, 244-45; free-standing, 27, 47, 51; grant facility for, 266; for institutional development, 12, 24-26, 89(13n), 174; lending operations, 71, 134
    Use of institutional twinning, 117, 118

Zaire
    Local consulting personnel, 115
    Technical cooperation to, 81

Zambia
    Capacity-building projects in, 25
    Free-standing TC projects in, 132
    TA personnel in: case study of, 22; as controllers, 95